2030

A BLUEPRINT FOR HUMANITY'S EXPONENTIAL LEAP

2030

A BLUEPRINT FOR HUMANITY'S EXPONENTIAL LEAP

ALAN SMITHSON
DOUG HOHULIN
HARVEY CASTRO, MD

UTP PUBLISHING SERVICES
UNIVERSITY OF TORONTO PRESS

Toronto Buffalo London

University of Toronto Press Publishing Services
Toronto Buffalo London
utorontopress.com

© Alan Smithson, Doug Hohulin, Harvey Castro, 2024

Library and Archives Canada Cataloguing in Publication

A catalogue record for this publication is available from Library and Archives Canada

ISBN   978-0-7727-1163-2 (paper)
ISBN   978-0-07727-1164-9 (EPUB)
ISBN   978-0-07727-1165-6 - (PDF)

Cover images: Designed using AI by Alan Smithson

We wish to acknowledge the land on which the University of Toronto Press oper-ates. This land is the traditional territory of the Wendat, the Anishnaabeg, the Haudenosaunee, the Métis, and the Mississaugas of the Credit First Nation.

UTP PUBLISHING SERVICES
UNIVERSITY OF TORONTO PRESS

DISCLAIMER

This book, *2030: A Blueprint for Humanity's Exponential Leap*, is designed for informational and educational purposes only. It is not intended to serve as a replacement for professional medical advice, diagnosis, or treatment. The authors, Alan Smithson, Dr. Harvey Castro, and Doug Hohulin, and the publisher have diligently worked to ensure that all information is accurate and up-to-date at the time of publication. However, medical practices, procedures, and the application of technology in healthcare continuously evolve, and they cannot guarantee the accuracy or applicability of any content at future dates.

Readers are encouraged to consult healthcare professionals for personal health or medical concerns. The author and publisher disclaim responsibility for any adverse effects or consequences from applying suggestions, products, or procedures discussed in this book.

The scenarios, case studies, and applications described are provided for illustrative purposes and, unless specifically mentioned, do not refer to any known individuals, locations, or situations. Any resemblance to actual persons, living or dead, or events is coincidental.

By using this book, you agree to assume all risks related to its use and hold the author and publisher harmless from any claims, losses, injuries, or damages arising from such use. Some of the content has been developed with the use of AI tools but the ideas, content, prompts and edits are from the authors.

DEDICATION

Alan Smithson

I dedicate this book to my wife Julie. Your love is precious and your patience is endless; you inspire me daily and I am proud to call you my best friend. To our two incredible daughters, Holly and Abi—you are the lights of my life, my greatest joy, and my deepest source of pride. Being your father is the greatest honour I could ever hope for. My love for you knows no bounds, and it is for your future, and the future of all children, that I write these words.

This book is born from the hope that those who read it will embrace a mindset of responsibility and possibility. That we will choose to protect this planet, our only home, and channel the astonishing power of technology to serve humanity first and foremost. My sincere wish is that we use these tools not for division, but for unity; not for exploitation, but for empowerment.

We are on the cusp of the exponential age, a time of extraordinary potential to build a brighter, fairer, and more sustainable world. If we get this right, we will create a future that resembles the visionary harmony of Star Trek—a world of exploration, progress, and shared purpose. But we must tread carefully, for the alternative is a Star Wars reality of conflict, division, and wasted potential.

And to you, the reader, the future is in your hands—may we rise to this moment with love, empathy, and hope. The choices we make today will define the legacy we leave for generations to come—let's get it right.

Doug Hohulin, B.Sc.EE

This book is dedicated to Byron Reese, my favourite author, who has inspired me to ponder and explore big ideas including the concept of "what your old," the risk of dying and the big history of the Infinite Progress and how humans come together to shape our world and our future in the Age of AI - the Augmented Agora. As Byron has written, may we use these tools of technology to "end ignorance, disease, poverty, hunger, and war."

May this work serve to inspire a future of billions of humans coming together to create a better world now and for future generations.

Harvey Castro, MD, MBA

This book is dedicated to my wife, whose love and support fuel my journey, and to my children, who inspire me to dream of a better world. To my mother, whose strength and values have shaped the person I am today, thank you for showing me what it means to lead with heart and purpose.

May this work ignite a shared vision where AI empowers humanity, not as a substitute for compassion but as a partner in progress. Together, we can heal, unite, and build a future defined by care, connection, and endless possibilities. Let's rise to this moment and leave a legacy that generations will be proud of.

Contents

Introduction

2030: A Blueprint for Humanity's Exponential Leap was written to share insights into the technological revolution that is upon us so that investors, executives, policymakers, educators, and everyday individuals are better informed and prepared for the exponential changes that will reshape our world in ways we can't yet fully comprehend or understand. *2030* aims to be an easy-to-read and easy-to-understand resource that will provide readers with tangible information, predictions and advice in hopes that they will take this information and leverage it in their daily work and personal lives to build a more sustainable, equitable, and beautiful world.

2030 explores the imminent wave of groundbreaking technologies, from quantum computing to Artificial Intelligence (AI), Artificial General Intelligence (AGI), Artificial Superintelligence (ASI), robotics, 3D printing, space travel, and climate change. This book delves into how these advancements will reshape our world, examining each topic through insightful questions and discussions. Each chapter will begin with a short story based on what will be possible in

2030, then explore a specific topic and offer a call to action. The chapter format is as follows:

1. Title
2. Relevant quote
3. Short story from 2030
4. Key predictions & stats
5. What is it?
6. What is going to happen?
7. Why does it matter?
8. Who will it impact?
9. Call to Action–One thing to be ready.
10. Expand on the topic (history, deeper explanation, etc.)
11. Recommended further reading

Throughout history, humanity has harnessed technology to improve lives, eradicate poverty, feed billions, travel to space, cure diseases, and communicate globally. This book continues that legacy, offering a comprehensive look at the future of innovation and its profound impact on our society.

Over the next five years, we will witness 50 years of transformation. Technologies seemingly unrelated to one another will merge to provide new value, create new products, and give us superpowers that we never thought possible earlier on. A combination of technologies, human endeavour, capitalism, and innovation are all coming together to expedite technology on a curve like we've never seen before. By 2030, the world will look very different than it does today.

We are on an exponential, accelerating curve defined almost perfectly by 'Moore's Law,' which states that the number of transistors on a microchip doubles approximately every two years, leading to exponential computing power and efficiency increases while reducing costs. Since the 1970s, this 'law' of exponential growth has held; still, from 2025-2030, we will see an acceleration of this exponential growth known as 'Huang's Law' named after NVIDIA CEO Jensen Huang, who states that advancements in graphics processing units (GPUs) are growing at a rate much faster than with traditional central processing units (CPUs). Huang's law states that the performance of GPUs will more than double every two years. Huang observes that NVIDIA's GPUs were "25 times faster than five years ago," whereas Moore's law would have expected only a ten-fold increase.

In addition to hardware, software advancements, such as artificial intelligence, are unleashing new capabilities and are compounding by improving hardware and software simultaneously. Semiconductors or 'chips' have continuously improved in accordance with Moore's Law.

> *"AI builds better chips, better chips power faster AI, which builds better chips that power faster AI in a virtuous cycle of innovation that accelerates endlessly."*

> **—Alan Smithson**

When thinking about the future, people tend to cite flying cars and space travel as their benchmark, but far more often the less glamorous technologies make a difference. Technologies like artificial intelligence (AI) have been relegated to science fiction or computer-significant discussions. Still, with the introduction of ChatGPT on November 30th, 2022, this technology has exploded into the cultural zeitgeist, putting the power of large language models (LLMs), discussed in the AI chapter of this book, into the hands of everyone everywhere. In addition to writing your kids' high school essays and helping you prepare that tedious report for work, these models are also solving complex problems such as protein folding, logistics and fraud detection, and also are solving seemingly less complex challenges across customer service, customized retail, marketing, translation and summarization. This swiss-army knife of technology has put incredible power into the hands of almost every human on earth with a promise to "elevate humanity," according to Sam Altman, CEO of OpenAI, whose stated mission is to build artificial general intelligence (AGI) defined as "highly autonomous systems that outperform humans at most economically valuable work," which "benefits all of humanity."

2030: A BLUEPRINT FOR HUMANITY'S EXPONENTIAL LEAP

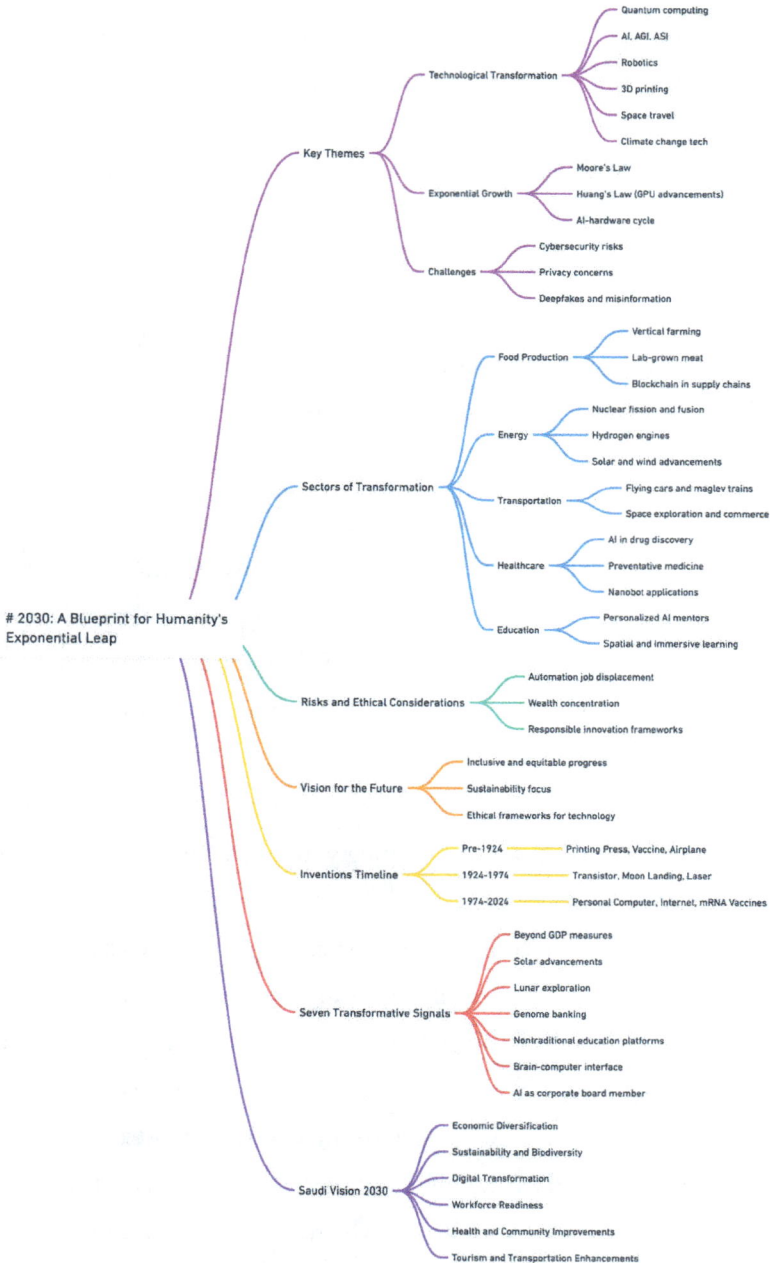

 # 2030: A Blueprint for Humanity's Exponential Leap

- **Key Themes**
  - Technological Transformation
    - Quantum computing
    - AI, AGI, ASI
    - Robotics
    - 3D printing
    - Space travel
    - Climate change tech
  - Exponential Growth
    - Moore's Law
    - Huang's Law (GPU advancements)
    - AI-hardware cycle
  - Challenges
    - Cybersecurity risks
    - Privacy concerns
    - Deepfakes and misinformation

- **Sectors of Transformation**
  - Food Production
    - Vertical farming
    - Lab-grown meat
    - Blockchain in supply chains
  - Energy
    - Nuclear fission and fusion
    - Hydrogen engines
    - Solar and wind advancements
  - Transportation
    - Flying cars and maglev trains
    - Space exploration and commerce
  - Healthcare
    - AI in drug discovery
    - Preventative medicine
    - Nanobot applications
  - Education
    - Personalized AI mentors
    - Spatial and immersive learning

- **Risks and Ethical Considerations**
  - Automation job displacement
  - Wealth concentration
  - Responsible innovation frameworks

- **Vision for the Future**
  - Inclusive and equitable progress
  - Sustainability focus
  - Ethical frameworks for technology

- **Inventions Timeline**
  - Pre-1924 — Printing Press, Vaccine, Airplane
  - 1924-1974 — Transistor, Moon Landing, Laser
  - 1974-2024 — Personal Computer, Internet, mRNA Vaccines

- **Seven Transformative Signals**
  - Beyond GDP measures
  - Solar advancements
  - Lunar exploration
  - Genome banking
  - Nontraditional education platforms
  - Brain-computer interface
  - AI as corporate board member

- **Saudi Vision 2030**
  - Economic Diversification
  - Sustainability and Biodiversity
  - Digital Transformation
  - Workforce Readiness
  - Health and Community Improvements
  - Tourism and Transportation Enhancements

The benefits of AI will come with significant challenges. Alongside its transformative potential, AI will introduce unique cybersecurity and privacy risks as systems become more interconnected and sensitive data increasingly relies on AI for management. The proliferation of deepfakes, capable of generating hyper-realistic but false images, videos, and audio, will blur the line between reality and fabrication, making it harder to trust what we see or hear. This will exacerbate issues of misinformation and disinformation, potentially fueling public distrust in media, governments, and institutions. Balancing the immense potential of AI with robust systems to counteract these risks will be critical to ensuring a future where trust, privacy, and integrity remain intact.

The field of robotics has come leaps and bounds (literally), as evidenced by the fun and playful videos of the robotics company Boston Dynamics. Elon Musk estimates that more than 10 billion humanoid robots will be deployed by 2040. These robots will be used for dangerous jobs like disaster rescue, nuclear reactor work, and mundane tasks such as house maintenance and factory assembly. We predict that there could be as many as one billion robots on Earth by 2033 to 2035, and if we consider Morgan Stanley's estimate of a $357 billion impact from eight million robots, scaling this up to one billion robots could theoretically result in an impact of over $44 trillion. However, this is likely an overestimate due to diminishing returns at scale. The chapter on the future of work explores this in more detail.

We stand at the precipice of an unlimited potential for humanity, but technology, a double-edged sword, is neither good nor bad; it is how we use it. For example, AI can find new drugs or create new pathogens; robots can help with manual labour or be turned into autonomous weapon systems. One risk that very few people are talking about is the risk to human capital when these autonomous systems start to replace human workers at scale.

According to global strategy and management consulting firm McKinsey & Company between 400-800 million jobs will be lost to AI and automation by 2030 (https://www.mckinsey.com/featured-insights/future-of-work/jobs-lost-jobs-gained-what-the-future-of-work-will-mean-for-jobs-skills-and-wages). An example of this will be eliminating virtually all human online customer service and support roles as AI is proven more empathetic, better at communicating, and drives more revenue at a much lower cost. This could mean a shorter work week for billions of people or a more severe concentration of wealth that is even more stark than what we see today, or likely both.

Technological advancements have compounding effects across fields such as agriculture, material science, healthcare, transportation, communication, energy, food, finance, and education. It is not unreasonable to assume that by 2030, students will have personalized AI mentors that allow them to hyper-accelerate their performance while creating new value almost instantaneously.

Advanced construction, manufacturing, and 3D printing technologies will transform how we build, offering pre-built modular homes and innovative designs that are more affordable, sustainable, and resilient to climate challenges. Materials like hempcrete, carbon-capturing concrete, and recycled composites will replace traditional materials, reducing carbon footprints and increasing durability. 3D printing will allow for rapid construction of homes and buildings with intricate, eco-friendly designs, minimizing waste and cutting costs. New structures will also integrate innovative technologies for energy efficiency and adaptability to changing climate conditions. These advancements will set a new standard for affordable, sustainable, and resilient architecture, making innovative housing accessible worldwide. We will use these technologies to create smarter cities with more green space and walking space designed to be carbon neutral.

Our energy consumption will skyrocket over the next five years, and we must find new, clean power sources. The energy revolution unfolding will be nothing short of transformative, a pivotal moment in human progress. By 2030, data centers alone will use 8-10% of the world's power, and the hyperscaler companies of today (NVIDIA, Google, Microsoft, Oracle, AWS) have begun building or purchasing nuclear reactors to run their data centers.

Nuclear fission and fusion, along with breakthroughs in antimatter energy, will redefine the boundaries of possibility. At the same time, hydrogen engines with built-in hydrogen generators run on water and produce O2 and water. Solar power, wind turbines, hydroelectric systems, and other clean energy sources will be further enhanced by novel energy storage technologies such as gravity and pressure batteries, propelling us toward a future of unprecedented abundance. This revolution promises near-limitless energy at nearly zero marginal cost, fundamentally altering industries and everyday life.

In addition, transportation will transcend its reliance on wheels, ushering in a new era of innovation. Flying cars, long the dream of futurists, will finally materialize as personal drones. As of this writing, Dubai has already begun construction on the world's first vertiport, with flights anticipated to commence as early as 2026, signalling the dawn of urban aerial mobility. Beyond the skies, maglev trains will reach unprecedented speeds, gliding effortlessly on magnetic tracks, while hypersonic planes promise to shrink global distances to mere hours.

The future of transportation will not be confined to Earth. Advances in propulsion systems, including ion drives, reusable rockets, and spaceplanes, will open the cosmos to exploration and commerce, connecting humanity to space and beyond. This transformation will redefine how we move, live, and dream, bridging continents and entire worlds. Humanity is expanding its presence in space, sending people, satellites, and mining capabilities beyond Earth's atmosphere. Space-based manufacturing facilities will leverage the unique conditions of microgravity to produce high-precision components, advanced materials, and even pharmaceuticals that are impossible to create on Earth.

Nanobot technology will revolutionize both manufacturing and healthcare. These microscopic machines will not only be used to construct new products with unparalleled precision but also operate inside the human body, curing diseases, repairing damaged tissues, and removing toxins at the cellular level. Nanobots will act as internal caretakers, enhancing health and extending lifespans in ways once thought impossible.

How we use the internet and media will fundamentally change. We will go from phone to face, from computer to simulator, and these technologies will not just be used for entertainment and gaming but for learning and education, marketing, retail, digital twins, collaboration, design, and counselling. No facet of humanity has been untouched by mobile phones, just like the coming wave of spatial computing. By 2030, there will be hundreds of millions of easy, light, and powerful heads-up displays, glasses, neural implants, and other technologies that give us a unique visualization of all the data that AI, AGI, and potentially ASI will generate in combination with more than eight billion humans connected to the internet.

We will preserve our sanity and enrich our lives through an ever-expanding array of entertainment media seamlessly integrated with the immersive capabilities of the spatial internet known as the 'metaverse' and powered by generative AI. Video, audio, books, and gaming will evolve into interactive, dynamic experiences that adapt to individual preferences. Generative AI will enable the creation of on-demand, custom content, allowing users to shape their narratives, worlds, and characters in real time. These advancements will transform entertainment into a highly personalized and participatory experience, fostering creativity, connection, and escapism while redefining how we consume and interact with media. This shift will make content infinitely diverse, ensuring there's always something uniquely engaging for everyone.

Blockchain technology may become the backbone of trust and transparency, quietly underpinning many aspects of our lives. While our digital assets and currencies may or may not exist on the blockchain, their role will extend far beyond finance. Blockchain will serve as a secure, immutable ledger for critical personal information—holding medical records, educational achievements, and digital asset ownership with unparalleled reliability. This technology will ensure that these essential components of our identities are easily accessible, universally verifiable, and protected from tampering, revolutionizing how we manage and share our most important data.

Food production will also transform profoundly, shifting from a global supply chain reliant on shipping to localized, technology-driven solutions. Lab-grown or "clean" meat will offer sustainable, cruelty-free protein, reducing the environmental impact of traditional livestock farming. Vertical farms, integrated into urban areas, will provide fresh produce year-round with minimal water and land use. 3D food printers will enable personalized, on-demand meals tailored to individual nutritional needs. This pivot toward local, tech-enabled food systems will redefine how we grow and consume food, minimizing waste, cutting transportation emissions, and ensuring fresher, healthier options for all.

By 2030, healthcare will undergo a profound transformation, shifting its focus from treating illnesses to proactively preventing them. With advancements in AI, synthetic biology, and groundbreaking tools like CRISPR, we will move closer to curing previously thought incurable diseases. AI will be pivotal in accelerating drug discovery and developing personalized treatment plans, tailoring preventative care to an individual's genetic makeup.

Innovations in synthetic biology will enable the creation of customized biological solutions, such as engineered cells and genetic edits, to eliminate hereditary diseases and combat illnesses before they manifest. Alongside these advancements, entirely new techniques, yet to be discovered, will expand our capabilities further. This shift to AI-driven, preventative healthcare will extend lifespans and significantly improve quality of life, ensuring a healthier and more resilient global population. Our chapter on when technology will save one million lives dives deeper into this concept.

The transformation will be more radical and impactful in education by 2030. The days of cramming information into our limited human brains will give way to adaptive, technology-driven learning, where knowledge is instantly accessible through smart devices and, soon, wearable technologies like augmented reality glasses. As computing evolves from flat screens to spatial computing, our phones will shift to face-worn interfaces, seamlessly integrating the digital and physical worlds. We will learn in new ways. We will attend classrooms remotely in virtual worlds. We will collaborate with people from all over the globe on pressing matters and our passions. Education will no longer be a linear path with an end goal but a lifelong pursuit of truth and learning.

In 2024, when this book was written, we witnessed the beginning of AI mentors, providing on-demand learning tailored to individual needs. These AI systems will teach and adapt in real-time, identifying strengths, addressing weaknesses, and guiding students through complex, multi-disciplinary challenges. Learning will move beyond memorization to focus on creativity, critical thinking, and hands-on problem-solving.

This future of education will equip learners with the tools to thrive in a world where knowledge is abundant, but applying it effectively will define success. It will mark the beginning of an era where everyone, regardless of location or background, has access to customized, immersive learning experiences that redefine education.

We are in a race—a relentless sprint toward the future where none of us can pause or step aside. It is imperative to use the insights in this book to make thoughtful, intentional decisions about why we are building these technologies, who they are for, and what outcomes we seek for humanity. The future is ours to shape, but we must do so with unbiased, safety and ethics at the forefront. These technologies—AI, nanobots, renewable energy, quantum computing, and more—hold the power to uplift society to unimaginable heights, yet they also carry the potential for destruction.

The next five years are pivotal. They must be dedicated to creating frameworks that ensure these innovations deliver opportunities for growth, exploration, and shared prosperity, enabling everyone to live an abundant life. This book provides a window into the transformative technologies redefining our world—each a remarkable achievement in its own right. Together, they represent the culmination of millennia of human endeavour, unlocking the actual superpowers of our age. Dive in and discover how we can shape this extraordinary future.

Inventions Prior to 1924

- Paper (105)
- Compass (12th Century)
- Printing Press (1440)
- Telescope (1608)
- Piano (1700)
- Steam Engine (1712)
- Refrigerator (1748)
- Gas Lighting (1792)
- Vaccine (1796)
- Battery (1800)
- Canned Food (1810)
- Stethoscope (1816)
- Bicycle (1817)
- Electromagnet (1825)
- Photography (1826)
- Matches (1826)
- Electric Motor (1832)
- Revolver (1836)
- Telegraph (1837)
- Saxophone (1840)
- Sewing Machine (1846)
- Anesthesia (1846)
- Glider (1853)
- Passenger Elevator (1857)
- Safety Pin (1849)
- Elevator (1853)
- Dynamite (1867)
- Typewriter (1868)
- Barbed Wire (1873)
- Jeans (1873)
- Telephone (1876)
- Phonograph (1877)
- Microphone (1877)
- Light Bulb (1879)
- Cash Register (1879)
- Electric Fan (1882)
- Roller Coaster (1884)

- Automobile (1886)
- Escalator (1891)
- Cinematograph (1892)
- Tractor (1892)
- Thermos (1892)
- Diesel Engine (1893)
- Zipper (1893)
- Radio (1895)
- X-ray Machine (1895)
- Radioactivity (1896)
- Aspirin (1899)
- Vacuum Cleaner (1901)
- Air Conditioning (1902)
- Airplane (1903)
- Plastic (1907)
- Washing Machine (1908)
- Model T Ford (1908)
- Assembly Line (1913)
- Stainless Steel (1913)
- Bra (1914)
- Tank (1916)
- Sonar (1917)
- Television (Mechanical, 1920)
- Traffic Light (1920)
- Pop-up Toaster (1921)
- Insulin (1922)
- Instant Camera (1923)

50 years of Inventions (1924 to 1974)

- Zipper (1923)
- Television (1927)
- Penicillin (1928)
- Jet Engine (1930)
- Radar (1935)
- Acrylic Fibers (1941)
- Nuclear Reactor (1942)
- Dialysis Machine (1943)
- Microwave Oven (1945)
- Tupperware (1946)
- Teflon Cookware (1946)
- Transistor (1947)
- Holography (1947)
- Crash Test Dummy (1949)
- Credit Card (1950)
- Birth Control Pill (1950)
- Super Glue (1951)
- Antidepressant (1952)
- Fiber Optics (1952)
- Scotchgard (1952)
- Heart-lung Machine (1953)
- Solar Cell (1954)
- Robots (1954)
- Lab-Grown Diamonds (1954)
- Velcro (1955)
- Hovercraft (1955)
- Polio Vaccine (1955)
- Video Cassette Recorder (VCR, 1956)
- Optical Fiber (1956)
- Satellite (1957)
- Fortran (Programming Language, 1957)
- Integrated Circuit (1958)
- Pacemaker (1958)
- CorningWare (1958)
- Flight Recorder (1958)
- Halogen Lamp (1959)
- Laser (1960)
- Space suit (1961)
- Soft contact lenses (1961)
- Carousel Slide Projector (1962)
- LED (1962)
- Valium (1963)
- Kevlar (1965)
- Moore's Law (1965)
- Dynamic Random Access Memory (DRAM, 1966)
- AI - Large Language Models (1966)
- Automated Teller Machine (1967)
- Handheld Calculator (1967)
- Floppy Disk (1967)
- VR/AR Research (1968)
- ARPANET (1969, precursor to the Internet)
- Boeing 747 (1969)
- Moon Landing (1969)
- Bank ATM (1969)
- AI - Natural Language Processing (1969)
- Floppy Disk (1971)
- CAT Scan (1971)
- Laser Printer (1971)
- Email (1971)
- Digital Watch (1972)
- Pong Video Game (1972)
- Mobile Phone (1973)
- Barcode (1974)
- 5.25 Inch Floppy Disk (1978)
- Apple II (1977)
- Compact Disc (CD, 1979)

50 years of Inventions (1974-2024)

- Personal Computer (1975)
- Digital Photography (1975)
- Magnetic Resonance Imaging (MRI) (1977)
- DNA Sequencing (1977)
- GPS (1978)
- Cellular Phone (1979)
- Sony Walkman (1980)
- Lithography (1980)
- 3D Printing (1981)
- Commodore Vic 20 (1981)
- MS-DOS (1981)
- Internet (TCP/IP) (1983)
- Touchscreens (1983)
- DNA Fingerprinting (1984)
- Flash Memory (1984)
- AI - Deep Learning (1986)
- Disposable Contact Lenses (1987)
- Autonomous Vehicle (1987)
- Modern Drone (1990)
- Gene Therapy (1990)
- World Wide Web (1991)
- Lithium-Ion Battery (1991)
- World Wide Web (1991)
- Smartphone (1992)
- Blue LED (1993)
- E-commerce (1994)
- Search Engine Optimization (1997)
- Wi-Fi (1997)
- Social Media (1997, early forms)
- Quantum Computing (1998)
- High-definition Television (1980s)
- Flash Memory (1984)
- Disposable Contact Lenses (1987)
- Digital Cellular Technology (GSM) (1991)
- Search Engine Optimization (1997)
- E-readers and E-Ink (1997)
- High-definition Television (1998)
- Vertical Farming (1999)
- Internet of Things (IoT) (1999)
- USB Flash Drive (2000)
- AI - Computer Vision (2000)
- Camera Phone (2000)
- Human Genome Project Completion (2003)
- Graphene (2004)
- Social Media Platform (2004)
- Video Streaming (2005)
- Video Streaming Service (2005)
- Cloud Computing (2006)
- Electric Car (Modern era) (2008)
- Smartphone App Ecosystem (2008)
- Blockchain Technology (2009)
- Mobile Wireless Charging (2009)
- Neural Interfaces (2010)
- Tablet Computer (2010)
- Smart Thermostat (2011)
- Virtual Assistants (2011)
- Solar Roof (2011)
- CRISPR Gene Editing (2012)
- Smart Watches (2013)
- Lab-Grown Meat (2013)
- Hyperloop (2013)
- Reusable Rocket (2015)
- Virtual Reality Headset (Commercial) (2016)
- Augmented Reality Headsets (2016)
- Smart Speaker (2014)
- Quantum Computer (2019)
- 5G (2019)
- mRNA Vaccines (2020)

- VR Meetings (2020)
- Tactile Virtual Reality (2020)
- AI Protein Folding Prediction (2020)
- Advanced AI-Language Models (2020)
- Autonomous Delivery Robots (2020)
- Tactile Virtual Reality (2020)
- Wireless Electricity (2020)
- Advanced AI for Earthquake Prediction (2020)
- Emotion Recognition AI (2020)
- Remote Surgery Systems (2020)
- Organic Light Emitting Diodes for Medical Applications (2020)
- Smart Prosthetics with Enhanced Sensory Feedback (2020)
- Autonomous Insect-sized Flying Robots (2020)
- AI-Enhanced Satellite Imaging for Climate Monitoring (2021)
- Wearable Air Purifiers (2021)
- Genome Editing for Allergen-Free Foods (2021)
- Large Scale Carbon Capture and Storage Facilities (2021)
- Advanced Smart Glasses (2021)
- AI-driven Algorithm for Predicting Protein Structures (2021)
- Self-repairing Roads (2021)
- Synthetic Genome (2021)
- High-Resolution Virtual Reality (2021)
- Molecular Coffee (2020)
- Graphene-enhanced Concrete (2021)
- Low Earth Orbit Satellite Internet (2021)
- Augmented Reality Surgical Systems (2021)
- Covid-19 Antiviral Pills (2021)
- Solar Windows (2021)
- Self-cleaning Materials (2021)
- Eco-friendly Hydrogen Production (2021)
- AI Drug Discovery (2021)
- CRISPR Diagnostic Tests (2021)
- James Webb Telescope (2021)
- Perseverance Rover (2021)
- Biodegradable Plastic Enzyme (2022)
- 6G Technology Demonstrations (2022)
- Advanced Quantum Computing Chips (2022)
- Lab-grown Wood (2022)
- Artificial Kidney (2022)
- Climate-Resistant Crops (2022)
- Digital Twins for Cities (2022)
- Synthetic Data for AI Training (2022)
- Voice-Activated AI Therapy Platforms (2022)
- AI-Driven Public Safety Networks (2022)
- Lunar Habitat Construction Technology (2022)
- Cognitive Enhancement Devices (2022)
- Virtual Reality Fitness Programs (2022)
- Energy-storing Bricks (2022)
- Neural Network AI for Autonomous Vehicles (2022)
- Blockchain for Secure Voting Systems (2022)
- AI-Powered Educational Bots (2022)
- Next-Generation Quantum Computers (2022)
- AI Systems for Real-time Language Translation (2022)

- Microbial Fuel Cells for Renewable Energy (2023)
- 3D Bioprinting of Human Organs for Transplant (2023)
- Space-based Solar Power Systems (2023)
- Personalized Medical Implants Using 3D Printing (2023)
- Smart Nanomaterials for Water Purification (2023)
- High-Performance Eco-friendly Insulation Materials (2023)
- Ultra-fast Broadband via Stratospheric Balloons (2023)
- Voice to Emotion AI for Mental Health (2023)
- Adaptive Traffic Control Systems (2023)
- AI Content Creation Tools (2023)
- Quantum Internet (2023)
- Bioengineered Rhino Horns (2023)
- Carbon Capture Technology (2023)
- Brain-Computer Interface for Enhanced Learning (2024)
- Longevity Drugs Targeting Aging (2024)
- Robotic Pollinators (2024)
- Gene Editing for Climate Adaptation in Crops (2024)
- Zero-Emission Hydrogen Trains (2024)
- Advanced Materials for Space Construction (2024)
- Quantum Encryption Technology (2024)
- Low-latency Satellite Internet (2024)
- AI-driven Predictive Healthcare Models (2024)
- Sustainable Aviation Fuel (2024)
- Biodegradable Electronics (2024)
- Water-from-Air Harvesting Systems (2024)
- Augmented Reality Interactive Learning Environments (2024)
- AI Optimization for Renewable Energy Systems (2024)
- Enhanced VR Immersion Techniques (2024)
- Personalized Nutrition AI Apps (2024)
- Autonomous Marine Research Drones (2024)
- Nanobot Drug Delivery Systems (2024)
- 3D Printed Sustainable Housing (2024)
- AI-Based Autonomous Security Systems (2024)
- Bioacoustic Monitoring for Wildlife Conservation (2024)
- Autonomous Urban Farming Systems (2024)
- Virtual Reality Home Design Tools (2024)
- Advanced Wearable Muscle Stimulators (2024)
- Deep Learning for Weather Prediction Enhancements (2024)
- AI for Real-time Traffic and Pollution Monitoring (2024)
- Precision Gene Editing for Coral Reef Restoration (2024)
- Ultra-Sensitive Biosensors for Virus Detection (2024)
- Smart Windows That Adjust to Light and Heat Automatically (2024)

Saudi Vision 2030 Key Focuses:

- Smoother Travel & Tourism
- Safer Roads, Less Accidents
- Protecting Biodiversity
- Access to Top Healthcare
- Healthier & More Active Lifestyles
- Home Ownership
- Inclusive Workforce / Women & Work
- Workforce Readiness
- National Investment Funds
- SME Support
- Economic Diversification
- Support Non-Profit Sector
- Improve Community
- Improve Food Development and Security
- Sustainable Water Usage
- Promote Financial Planning
- Privatize Government Services
- Build Advanced Capital Markets
- Attract Foreign Investment
- Digitizing Government Services
- Health Crisis Preparedness
- Increase Fee Revenue without Increasing Taxes
- Maximize Oil Revenue
- Leaner & More Effective Government
- Enhanced transparency of government
- Increase the quality of services for citizens
- Develop e-Government

APPENDIX:

Seven transformative signals for the future of humanity

Beyond GDP: Redefining Progress: The world is moving beyond GDP to measure societal well-being, sustainability, and human flourishing.

1. **Harnessing Solar Potential**: Advances in solar technology make doubling global energy from the sun an achievable milestone addressing the climate crisis.

2. **Returning to the Moon**: Humanity's return to the moon rekindles curiosity and inspires new frontiers of innovation.

3. **A Genome Bank of One Billion Samples**: A genome bank promises breakthroughs in personalized medicine, advancing global health.

4. **Learning Outside School**: Five million students now access education through innovative, nontraditional platforms.

5. **Brain-Computer Chip in a Healthy Person**: The first brain-computer chip implant signals a leap in neuroscience and human-machine interaction.

6. **AI Board Member in a Fortune 500 Company**: An AI entity now sits on a Fortune 500 board, redefining strategic corporate decision-making.

Food

George stood at the edge of his sprawling farm in California, surveying a landscape transformed by technology. The traditional rows of strawberries, lettuce, and tomatoes stretched before him. Still, now they were bordered by sleek, vertical farming towers—innovative structures growing crops stacked layer upon layer, sky-high. These vertical farms allowed George to cultivate year-round despite the droughts or heat waves that sometimes swept through California. Inside the towers, an AI-controlled climate system optimized the light, humidity, and nutrient mix, producing fresh vegetables with less water and fewer resources.

George's farm wasn't just a marvel of vertical growth; it was a scene out of a sci-fi novel, thanks to AI-driven laser weed control. Instead of spraying fields with pesticides, AI-guided laser beams targeted only invasive weeds, leaving his crops untouched and healthier than ever. Across the way, his new food bioreactor was a sight he hadn't imagined just a few years ago. The sleek machine produced lab-grown proteins, adding a futuristic twist to his otherwise traditional farming methods. He saw it as both a nod to the future and a new way to supply his growing roster of clients around the world with sustainable protein alternatives.

Meanwhile, thousands of miles away in Dubai, Lucy had an equally transformative experience as a food importer and exporter. Sitting at her high-tech workspace, she checked her AI-optimized logistics dashboard, where worldwide shipments were tracked down to the minute. Every item George has sent her—from tomatoes to strawberries—was meticulously timed and tracked with blockchain verification to guarantee peak freshness upon arrival. She remembered back in 2024 when nearly a third of all food produced globally was wasted before reaching consumers due to inefficiencies in the supply chain. Now, thanks to AI-driven logistics and predictive algorithms, her operation has nearly eliminated that waste, with AI handling inventory predictions and arranging shipments for optimal delivery times.

It was Friday evening when Lucy and George hopped on a video call, each eager to exchange stories from their tech-driven food worlds. "George, you wouldn't believe it," Lucy began, excitedly leaning in. "We've just started using 3D food printers here in Dubai to create personalized meals for customers. We're blending ingredients from all over the world to craft dishes on demand. And your strawberries? They're a favourite!"

George laughed, genuinely impressed. "Well, Lucy, over here, I've got autonomous drones flying over the fields, making deliveries to the distribution hubs. They're like little worker bees, taking the crops straight from the towers and bioreactor to the transport trucks. No fuel, no wasted time, and reduced labour costs. The drones have been a game changer for us."

Lucy's eyes sparkled as she imagined George's drones flitting about his fields like futuristic farmhands. "It's incredible, George. The food system has changed so much. Remember all that spoilage we used to have? Between blockchain, AI logistics, and precise demand tracking, I feel like we're finally bringing fresh, quality food to every table."

They both momentarily fell silent, realizing how far their worlds had evolved. Not only had food waste drastically decreased, but so had their reliance on traditional methods that once strained the environment. Lucy, with her focus on logistics and consumer satisfaction, and George, with his commitment to sustainable farming, represented a new era in food production and distribution.

"We're living in a world," George said, "where I can grow more food on less land, with less water, and still reach people across the globe. And it feels like we're doing it right, with no harm to the soil or waste."

Lucy nodded. "Yes, and I can get that food to the farthest parts of the world with AI coordinating every step. It's all so... connected now."

As they said goodbye, George looked over his farm, imagining all the meals his strawberries, tomatoes, and lab-grown proteins would soon create in Dubai. Lucy, closing her laptop, envisioned a future where food waste was a thing of the past, and every bite was fresh, sustainable, and carefully curated. In this new age of AI, vertical farming, and ultra-efficient logistics, they were not just part of the food chain but actively shaping a more sustainable, abundant future.

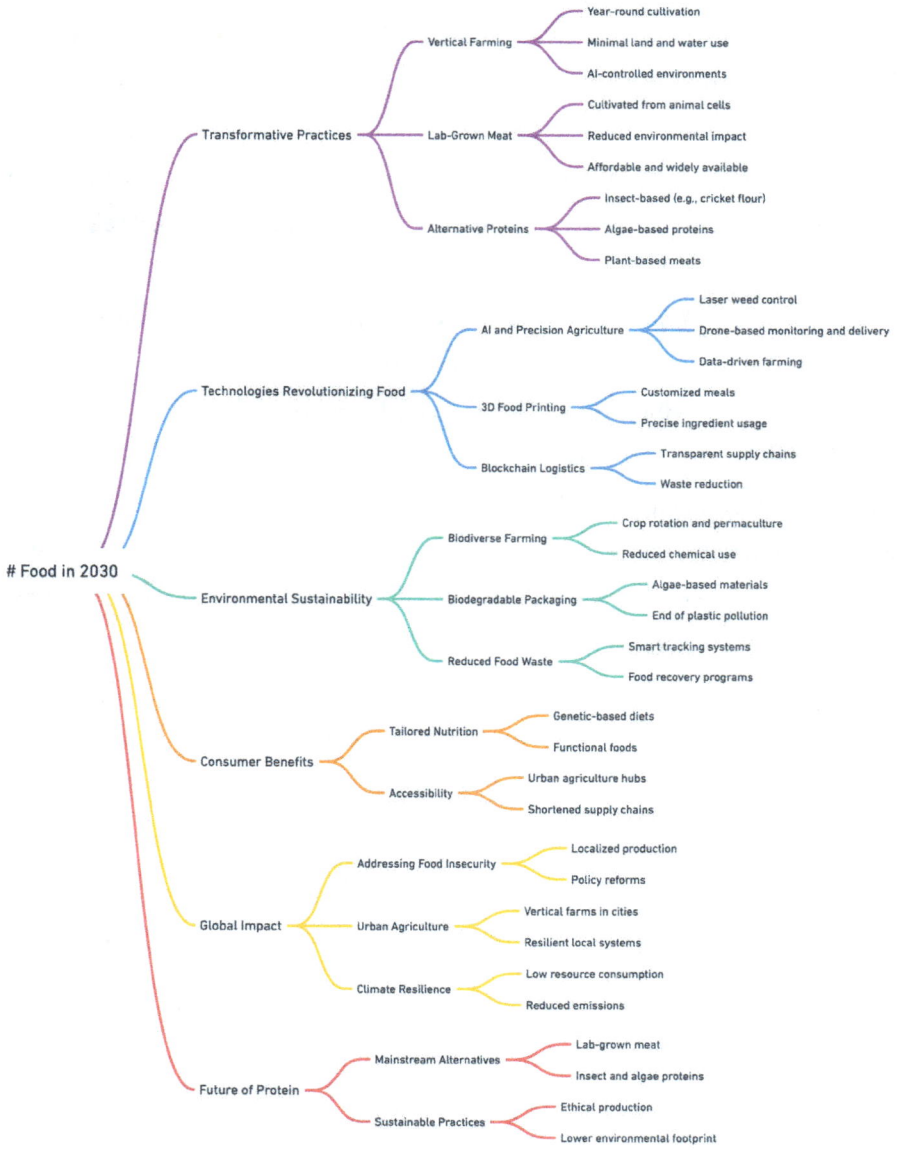

FOOD IN 2030

 # Food in 2030

- **Transformative Practices**
  - **Vertical Farming**
    - Year-round cultivation
    - Minimal land and water use
    - AI-controlled environments
  - **Lab-Grown Meat**
    - Cultivated from animal cells
    - Reduced environmental impact
    - Affordable and widely available
  - **Alternative Proteins**
    - Insect-based (e.g., cricket flour)
    - Algae-based proteins
    - Plant-based meats

- **Technologies Revolutionizing Food**
  - **AI and Precision Agriculture**
    - Laser weed control
    - Drone-based monitoring and delivery
    - Data-driven farming
  - **3D Food Printing**
    - Customized meals
    - Precise ingredient usage
  - **Blockchain Logistics**
    - Transparent supply chains
    - Waste reduction

- **Environmental Sustainability**
  - **Biodiverse Farming**
    - Crop rotation and permaculture
    - Reduced chemical use
  - **Biodegradable Packaging**
    - Algae-based materials
    - End of plastic pollution
  - **Reduced Food Waste**
    - Smart tracking systems
    - Food recovery programs

- **Consumer Benefits**
  - **Tailored Nutrition**
    - Genetic-based diets
    - Functional foods
  - **Accessibility**
    - Urban agriculture hubs
    - Shortened supply chains

- **Global Impact**
  - **Addressing Food Insecurity**
    - Localized production
    - Policy reforms
  - **Urban Agriculture**
    - Vertical farms in cities
    - Resilient local systems
  - **Climate Resilience**
    - Low resource consumption
    - Reduced emissions

- **Future of Protein**
  - **Mainstream Alternatives**
    - Lab-grown meat
    - Insect and algae proteins
  - **Sustainable Practices**
    - Ethical production
    - Lower environmental footprint

WHAT IS IT?

The food landscape will be a fascinating blend of advanced technology and sustainable practices, creating an efficient, environmentally conscious system tailored to individual health needs. Lab-grown meats and plant-based alternatives will be readily available, providing protein options that mimic the texture and flavour of traditional meats without the environmental impact of livestock farming. 3D-printed foods will allow for customizable meals that cater to unique dietary needs, with nutrients and flavours adjusted for each individual.

Vertical farms will rise within urban centers, using minimal land and water to grow fresh produce year-round. These vertical farming systems will utilize precision agriculture, relying on AI and data-driven tools to monitor soil health, optimize water usage, and control plant growth, all while reducing pesticide use. Advances in biotechnology will create nutrient-dense, resilient crops, further increasing efficiency and reducing resource use in agriculture.

Efforts to combat food waste will be integral to the food system, with AI-powered logistics helping to track and manage supply chains, ensuring food is used before it spoils. Additionally, alternative proteins such as algae, insect-based proteins, and cultured meat will gain mainstream acceptance as sustainable, nutrient-rich options. These proteins offer a lower environmental footprint, promoting a balanced global diet and addressing food scarcity.

Food production, logistics and preparation in 2030 will be innovative and sustainable, shaped by technology, and driven by a commitment to health and environmental responsibility. This future of food will cater to diverse tastes and nutritional needs while prioritizing planet-friendly production methods that make fresh, customized, and sustainable food accessible to everyone.

WHAT IS GOING TO HAPPEN?

Food production will be characterized by localized, sustainable practices prioritizing environmental health and community access to fresh food. Vertical and urban agriculture will transform cityscapes, enabling neighbourhoods to enjoy locally grown produce year-round. This will provide communities with fresher food options and significantly reduces transportation emissions, lowering the overall carbon footprint of food distribution.

Our protein sources will shift significantly as alternative proteins become mainstream. Lab-grown meats, cultivated from animal cells without the need for raising and slaughtering livestock, will offer sustainable and ethical options that closely mimic traditional meats in taste and texture. Simultaneously, insect protein will gain acceptance as a highly efficient and nutritious food source. Insects, like crickets and mealworms, are rich in protein, vitamins, and minerals and require far fewer farm resources than conventional livestock. Incorporating insect protein into foods—whole or processed into flours and powders—will become commonplace, contributing to food security and environmental sustainability. Additionally, algae-based proteins will provide yet another option, particularly valued for their high nutritional content and minimal environmental impact, adding diversity to sustainable protein sources.

In tandem, 3D printing technology will revolutionize meal preparation, allowing individuals to create on-demand meals tailored to their specific nutritional needs and taste preferences. This technology will seamlessly integrate alternative proteins into familiar food formats, making the transition to new protein sources more palatable for consumers.

Precision farming, powered by AI, drones, and data analytics, will enhance crop yields with unprecedented accuracy. Drones will survey fields, monitor soil health, and assess crop conditions, allowing farmers to apply water, nutrients, and pest control measures precisely where needed, minimizing waste and conserving resources. These practices support eco-friendly farming and lead to healthier soil, reduced water usage, and lower chemical inputs.

Overall, these advancements will create a more resilient and adaptable food system capable of meeting the nutritional needs of a growing global population while minimizing environmental impact. By embracing alternative proteins and sustainable technologies, food production in 2030 will feed people in ways that preserve the planet and promote global health.

WHY DOES IT MATTER?

With the global population expected to exceed 8.5 billion, more than traditional farming methods will be needed to sustain food demand. The future of food must address critical challenges like climate change, resource scarcity, and food security. Innovations in food production will reduce the environmental impact of agriculture, mitigate the risks of food shortages, and provide healthier, personalized options for consumers. Furthermore, reducing reliance on livestock farming will decrease greenhouse gas emissions, improve animal welfare, and conserve valuable resources such as land and water.

Traditional farming methods heavily affect our environment—they consume vast amounts of water, require expansive land, emit significant greenhouse gasses, and generate substantial waste. With climate change accelerating and natural resources dwindling, we must rethink how we produce and consume food.

Innovations empower us to address food security in ways we've never been able to before. We can maximize crop yields while conserving water providing resilience against climate shocks and supply chain disruptions. With 3D food printing and personalized nutrition, we're moving toward a future where everyone can access food tailored to their needs, reducing waste and promoting health.

Ultimately, these innovations represent a commitment to a future where food production is sustainable, equitable, and resilient. They are the building blocks of a food system that respects both people and the planet, ensuring that we can nourish generations to come without compromising the environment or global stability.

This is why rethinking food production isn't just important—it's essential for a sustainable and thriving future.

WHO WILL IT IMPACT?

Advancements in food technologies will touch every part of society, transforming the lives of farmers, food producers, consumers, urban populations, policymakers, and the food industry itself. For farmers, emerging technologies like AI-guided precision agriculture, automated machinery, and crop bioreactors will improve yields, conserve resources, and reduce costs. These tools will allow farmers to grow food more efficiently while being resilient to climate pressures, ensuring stable incomes and more sustainable practices. Vertical farms in urban centers will become a powerful tool, enabling cities to grow produce locally, cut transportation emissions, and increase food security in densely populated areas.

Consumers will benefit from a wide range of new options. Alternative proteins like lab-grown meats, insect-based foods, and algae will bring protein sources that are healthier, more sustainable, and more ethically produced. Functional foods fortified with added nutrients or tailored to boost immunity, digestion, and mental health will be available to support well-being at a personal level. Genetic-based nutrition plans will also offer dietary recommendations based on individual genetic profiles, helping people make choices aligned with their unique needs. With 3D food printing, consumers will have the ability to create customized meals tailored to specific nutritional requirements and taste preferences, reducing food waste, and personalizing nutrition in a way that was unimaginable in the past.

Plant-based proteins, such as tofu, tempeh, and seitan, have long been a staple of vegetarian and vegan diets. However, the rise of plant-based proteins is broader than just traditional plant-based products.

Urban populations will enjoy unprecedented access to fresh produce and specialized foods thanks to urban agriculture and localized tech-driven food hubs. City dwellers will find it easier to access fresh food, grown nearby, as vertical farms and AI-driven greenhouses become integral parts of urban infrastructure. Policymakers will face the challenge of adapting regulations to keep pace with these innovations, creating frameworks to manage lab-grown meat, alternative proteins, genetic-based nutrition products, and other emerging technologies safely and equitably. They must also promote sustainable farming and local food production to ensure community accessibility.

The global food industry will be transformed as supply chains become shorter, more technologically integrated, and more efficient. Blockchain will enable transparent, traceable supply chains that reduce food fraud, improve safety, and enhance consumer trust. As food waste reduction technologies powered by AI logistics streamline storage and distribution, the industry can supply fresher food more efficiently and sustainably.

Overall, this revolution in food technology and sustainability will create a healthier, more resilient, and adaptable system to a growing population's needs, making nutritious, eco-conscious food more accessible worldwide. The impact will extend across industries, redefining what it means to grow, access, and consume food in a world shaped by innovation and sustainability.

CALL TO ACTION

To prepare for the food revolution of 2030, governments, industries, and educational institutions must prioritize investments in research, infrastructure, and workforce training that support sustainable food technologies. CEOs can start by embracing sustainable food innovations within their companies and integrating alternative proteins and sustainable sourcing practices to reduce environmental impact and appeal to conscious consumers. Investors should consider allocating funds toward emerging food tech sectors, such as precision agriculture, lab-grown meats, and genetic-based nutrition solutions, positioning themselves at the forefront of this evolving market. Government bureaucrats can focus on creating policies and incentives that facilitate the adoption of new technologies, from vertical farming tax credits to safety regulations for alternative proteins. To ensure a smooth transition, all sectors should work on educating the public about these new foods and technologies, encouraging awareness and acceptance to build a sustainable, secure food future.

Cultured Agriculture and Alternative Proteins in 2030

By 2030, **cultured agriculture** will be a significant part of the global food supply, encompassing various alternative proteins like insect-based products, lab-grown meat, and plant-based alternatives. Insects, such as crickets and mealworms, will be mass-farmed for their high protein yield and minimal environmental footprint, offering a sustainable protein source with reduced resource demands. Lab-grown meat, cultivated from animal cells without traditional livestock farming is an affordable alternative that provides the taste and texture of real beef with a fraction of the environmental impact. **Plant-based proteins,** made from soy, pea protein, and fungi, will continue to thrive, driven by eco-conscious consumers and the growing demand for healthier, ethical food choices. These innovations in cultured agriculture will significantly reduce the strain on land, water, and biodiversity while helping to meet the dietary needs of a growing global population.

By 2030, alternative proteins will be a significant part of the global diet, including insect-based products, lab-grown "clean" meat, and plant-based alternatives. Insects like crickets will be farmed for their high protein content and low environmental impact. At the same time, lab-grown meat (produced from animal cells without slaughter) will be widely available and affordable. Plant-based meats, made from soy, pea protein, and other ingredients, will continue to grow in popularity, driven by environmental concerns and health-conscious consumers. These alternatives will help reduce the environmental strain caused by traditional livestock farming and offer a more sustainable way to feed the growing global population.

Back to Truly Organic, Biodiverse Crop Sources

In 2030, agriculture will return to more **organic, biodiverse farming** methods. Instead of relying on monoculture, which depletes the soil and harms ecosystems, farmers will cultivate various crops that improve soil health and promote natural pest control. Biodiversity farming will produce nutrient-rich crops while reducing the need for chemical fertilizers and pesticides. Regenerative agriculture techniques, including crop rotation, permaculture, and agroforestry, will help restore ecosystems and fight climate change by sequestering carbon in the soil. Consumers will have greater access to organic, sustainably grown foods, creating healthier diets and ecosystems.

Local, Smaller Producers

By 2030, **local and smaller food producers** will play a more significant role in feeding communities. With advancements in vertical farming, aquaponics, and indoor farming, local producers can grow fresh produce year-round, even in urban environments. This shift towards local production will reduce the environmental footprint of transporting food long distances and support more resilient local food systems. Farmers' markets, community-supported agriculture (CSA), and direct-to-consumer platforms will flourish, giving consumers fresher, healthier, and more sustainable food choices.

Blockchain Tracking of Ingredients

In 2030, **blockchain technology** will be widely used to track the provenance of ingredients from farm to table. Blockchain's immutable ledgers will provide transparency across the food supply chain, allowing consumers to verify the origin, quality, and sustainability of the food they buy. This will help prevent food fraud, ensure fair trade, and allow producers to showcase the ethical practices behind their products. Blockchain will also enhance food safety, making it easier to trace and recall contaminated products in the event of foodborne illness outbreaks.

Logistics

In 2030, food transportation logistics will be highly efficient, leveraging **AI, IoT, and blockchain** to streamline supply chains. Automated systems will manage inventory, track shipments in real time, and optimize delivery routes to reduce fuel consumption and spoilage. Drones and autonomous vehicles will be used to deliver food faster, particularly in urban and remote areas. Cold chain technology will ensure that perishable goods remain fresh during transport, reducing food waste. These innovations in logistics will create faster, more sustainable food distribution systems, and ensure that food reaches consumers in peak condition.

Addressing Food Waste

By 2030, technology and policy measures will dramatically reduce food waste. AI systems will help optimize food production, preventing overproduction, while **innovative packaging** and food tracking will alert consumers and retailers when products are nearing expiration. Food waste recovery programs will also expand, repurposing surplus food for animal feed, compost, or energy generation. Governments will enforce stricter regulations on food waste management, and consumer education campaigns will raise awareness about food waste's environmental and economic impact.

Food Insecurity

By 2030, **food insecurity** will be tackled through **sustainable agriculture, efficient distribution systems,** and policy reforms. Advanced farming techniques like vertical and precision agriculture will increase food production without depleting natural resources. Governments and international organizations will invest in infrastructure to ensure food reaches underserved communities. Digital platforms will connect farmers directly with consumers, reducing waste and lowering costs. Public-private partnerships will address systemic barriers to food access, ensuring that nutritious and affordable food is available.

Fast-Food Restaurants and the Future of Beef

By 2030, fast-food giants will face significant changes as consumer preferences shift toward sustainable and ethical food sources. In response to environmental concerns, fast-food chains will reduce their reliance on traditional beef, replacing it with **plant-based or lab-grown alternatives**. These shifts will help reduce the environmental impact of beef production, including methane emissions and deforestation. Restaurants will also innovate with eco-friendly packaging and more sustainable business practices, aligning with consumer demands for more ethical and environmentally responsible food options.

China, Brazil, the Rainforest, and Better Meat Sources

By 2030, efforts to preserve the **Amazon Rainforest** in Brazil will intensify as deforestation for cattle ranching becomes unsustainable. China and Brazil, two of the largest meat producers and consumers, will shift toward **sustainable livestock practices** and **alternative protein sources**. This transition will be driven by international pressure and domestic policies to reduce the environmental impact of meat production. Sustainable ranching methods, reforestation projects, and a shift toward plant-based and lab-grown meat will help protect the Amazon while meeting the growing global demand for protein.

3D Food Printing

By 2030, **3D food printing** will revolutionize how we prepare and consume food. Using edible pastes made from ingredients like vegetables, proteins, and carbohydrates, 3D printers will create customized, nutritious meals on demand. This technology will be valuable in restaurants, hospitals, and space travel, where precision nutrition and convenience are critical. Consumers can print their meals at home, tailoring the ingredients and portion sizes to their dietary needs. 3D printing will reduce food waste by allowing for precise ingredient usage and will help make meal preparation faster and more efficient.

Food Production Automation Starts

By 2030, **automation in food production** will be widespread, with robots and AI systems managing everything from planting and harvesting to sorting and packaging. Autonomous tractors, drones, and harvesting machines will work together to optimize crop yields, while AI will analyze data to predict the best planting times and resource allocation. Automated factories will process, package, and ship food with minimal human intervention, improving efficiency and reducing labour costs. This shift will help meet the growing demand for food while making agriculture more sustainable and less reliant on manual labour.

Autonomous Food Growth, Preparation, and Delivery

In 2030, **autonomous systems** will oversee food production and consumption, from growing crops to delivering meals. **Autonomous farms** will use AI and robots to plant, water, and harvest crops with minimal human intervention. At the same time, **smart kitchens** will prepare food automatically based on dietary preferences and health data. **Autonomous delivery vehicles** and drones will bring groceries and prepared meals directly to consumers, reducing delivery times and transportation costs. This end-to-end automation will revolutionize how food is produced, prepared, and consumed, offering unparalleled convenience and efficiency.

Biodegradable Packaging (The End of Plastic)

By 2030, **biodegradable packaging** will replace traditional plastic in the food industry, dramatically reducing pollution and waste. Made from materials like algae, seaweed, and plant-based polymers, biodegradable packaging will decompose naturally without leaving harmful residues. This shift will be driven by both government regulations and consumer demand for more sustainable products. Biodegradable packaging will help address the global plastic crisis and create a circular economy where packaging materials are composted or recycled, reducing their environmental footprint.

Alternative Proteins

The need for alternative proteins is clear. Animal agriculture significantly contributes to greenhouse gas emissions, deforestation, and water pollution. The livestock industry is also a major driver of biodiversity loss, with many species facing extinction due to habitat destruction and fragmentation. Furthermore, meat production significantly contributes to global food waste, with an estimated one-third of all food produced globally being lost or wasted. In response, alternative protein sources offer consumers a more sustainable and environmentally friendly option.

Insect-Based Proteins

Insects are one of the most promising alternative protein sources. Insect-based proteins, such as cricket flour and mealworms, are rich in protein, fibre, and micronutrients. Insect farming is more sustainable than traditional livestock production, requiring significantly less land, water, and feed. Insect-based proteins are already used in various applications, from animal feed to human nutrition. Companies like Six Foods and Tiny Farms are pioneering the development of insect-based protein products, including snack bars, protein powders, and even insect-based meat alternatives.

Lab-Grown Meat

Lab-grown meat, or clean meat, is another innovative alternative to traditional meat. By 2030, **lab-grown meat** (cultured or clean meat) will be a mainstream alternative to traditional livestock farming. Lab-grown meat is created by culturing animal cells in a controlled environment, eliminating the need for animal slaughter and reducing the environmental impact of meat production. It will closely mimic conventional meat's taste, texture, and nutritional profile without the environmental, ethical, or health issues associated with industrial animal farming. Cultured meat will significantly reduce land, water, and feed requirements while eliminating the need for antibiotics and reducing greenhouse gas emissions. This innovation will offer a scalable and sustainable way to meet global meat demand while improving food security. Companies like Memphis Meats and Mosa Meat are leading the charge in lab-grown meat development, with products already being tested in restaurants and stores. Lab-grown meat offers a more sustainable and humane option for consumers, potentially reducing greenhouse gas emissions by up to 75%.

The Future of Alternative Proteins

As the alternative protein market continues to evolve, it's clear that the future of food is not just about what we eat but how we produce it. The rise of alternative proteins is about reducing environmental impact and creating a more sustainable and equitable food system. As consumers become increasingly aware of their food choices environmental and social implications, the demand for alternative proteins will likely continue to grow.

Healthcare: A New Era of Medical Transformation

"The true revolution of healthcare is not driven by machines, data, or algorithms, but by the extraordinary potential of AI to amplify what makes us human—our empathy, creativity, and drive to care for one another. In the end, technology is not the solution but the enabler of deeper, more meaningful human connections in medicine."—**Harvey Castro, MD**

Beverly Hart had spent her life listening. Patients confided in her about their aches, fears, and unspoken worries. She was the trusted family doctor in a small suburban practice, her office a warm haven filled with soft lighting and a wall of patient thank-you notes from over the years. But beneath her caring demeanour, Beverly carried a quiet exhaustion. The countless hours she spent each week typing notes and handling a barrage of communications had dulled her joy in practicing medicine. She was always there for her patients, but who, she wondered, was there for her?

Meanwhile, Alvin Reyes was crafting a revolution in his downtown Toronto tech lab, though he wouldn't call it that. "I'm just solving a problem," he'd say with a grin. Alvin was a sharp, inventive software engineer who had seen firsthand how fragmented healthcare could be when his mother missed a critical follow-up due to confusing medical records. He vowed to create an AI-powered assistant to capture every word of a doctor's visit, organize it into actionable insights, and streamline medical records. His system, "MedMosaic," was his masterpiece, but to Alvin, it wasn't finished. He needed someone in the trenches who could help him refine it—a true partner who understood the heart of healthcare.

One rainy Friday afternoon, Beverly attended a healthcare innovation conference at the insistence of a colleague. She wasn't one for tech talks—words like "blockchain" and "machine learning" made her head spin—but something about a session titled "Reclaiming the Art of Medicine Through AI" caught her eye. She sat near the back, sipping lukewarm coffee, as a young man with dishevelled hair and boundless energy took the stage.

Alvin didn't use slides. He spoke directly to the crowd, weaving a story about his mom's missed appointment and the chaos of her medical records. "Doctors are drowning," he said passionately. "Not in medicine, but in paperwork. What if we could give you back your time and focus so you could do what you do best?"

Beverly felt her skepticism melting away. She approached Alvin after the talk. "I'm not a 'tech person,'" she began, "but I think you're onto something."

What began as a brief conversation turned into weeks of collaboration. Beverly tested MedMosaic in her practice, providing Alvin with detailed feedback. At first, the AI assistant struggled to keep up with the nuances of medical conversations. It miscategorized symptoms, bungled follow-up notes, and occasionally sent Beverly's patients reminders in Comic Sans font. But Alvin was relentless. He spent late nights tweaking algorithms and reprogramming features based on Beverly's insights.

And Beverly? She discovered a spark of excitement she hadn't felt in years. She loved seeing the system improve and loved the idea that she was shaping something that could help doctors like her. She started telling Alvin stories about her patients—their medical histories and their lives. Alvin listened intently, using her anecdotes to make the AI more intuitive and empathetic.

One day, Alvin dropped by Beverly's clinic to observe MedMosaic. It was a busy morning, and Beverly was behind schedule. A patient named Helen, a retired teacher with a history of heart issues, arrived for her annual check-up. Beverly visited as usual, but something about Helen's symptoms seemed... off. MedMosaic flagged inconsistencies in Helen's recent vitals and suggested further tests.

"Helen," Beverly said gently, "I'd like to run some additional bloodwork."

The results revealed early signs of a serious heart condition that might have been missed without the AI's keen eye. Helen hugged Beverly tightly after the diagnosis. "You always take care of me," she said, tears in her eyes.

Later that day, Beverly found Alvin sitting in her office, staring at his laptop with a sheepish grin. "Did it work?" he asked.

Beverly didn't answer right away. Instead, she sat beside him, looking at the screen where the system's algorithms had analyzed Helen's case. "You didn't just make a tool," she said finally. "You gave me my instincts back."

As MedMosaic rolled out to more clinics, Beverly and Alvin became inseparable allies in their mission to transform healthcare. Their lives intertwined in unexpected ways. Beverly was drawn into Alvin's tech world, attending late-night hackathons and brainstorming sessions. She even started carrying a notebook to jot down patient anecdotes for Alvin's endless quest to improve the system's empathy.

Alvin learned the art of listening from Beverly. He started joining her on early morning walks with her golden retriever, Max, where she shared her frustrations and hopes for the future of medicine. Those quiet moments grounded Alvin, reminding him that the heart of his work wasn't code—it was people.

Years later, Beverly and Alvin stood on stage at an international healthcare summit. They had come a long way from that conference on a rainy afternoon. MedMosaic was now a cornerstone of modern medical practices, used in clinics and hospitals worldwide. Alvin spoke about the technical innovations behind the system, but Beverly brought the room to its feet.

"Medicine isn't just science," she said, her voice steady. "It's a relationship. And what Alvin has created doesn't replace that—it amplifies it. Because of him, I can listen to my patients again. I can see them. And isn't that what healing is all about?"

Back in her clinic, Beverly still kept her wall of thank-you notes. But now, one from Alvin was pinned right in the center. It simply read:

"For showing me what care truly means."

And every time Beverly saw it, she smiled, knowing their story had just begun.

HEALTHCARE IN 2030

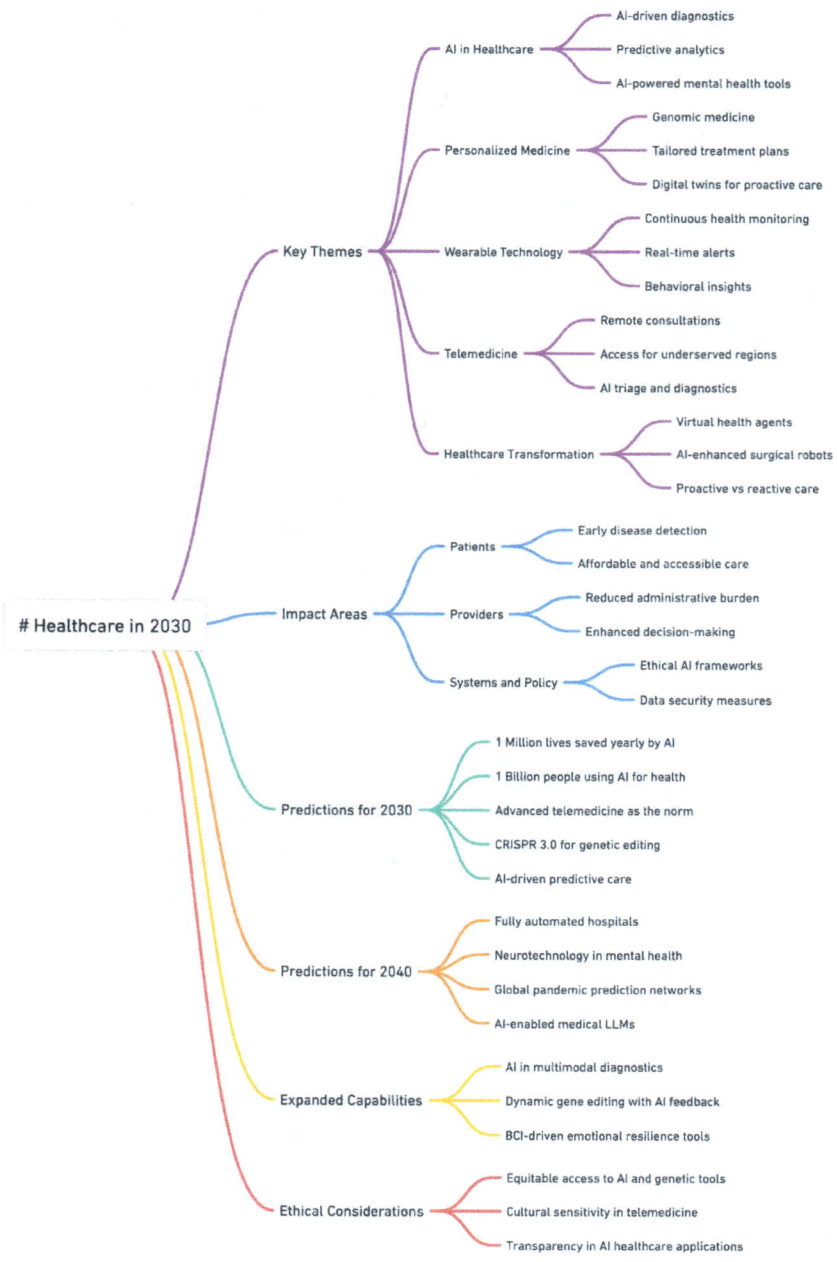

- **# Healthcare in 2030**
  - **Key Themes**
    - AI in Healthcare
      - AI-driven diagnostics
      - Predictive analytics
      - AI-powered mental health tools
    - Personalized Medicine
      - Genomic medicine
      - Tailored treatment plans
      - Digital twins for proactive care
    - Wearable Technology
      - Continuous health monitoring
      - Real-time alerts
      - Behavioral insights
    - Telemedicine
      - Remote consultations
      - Access for underserved regions
      - AI triage and diagnostics
    - Healthcare Transformation
      - Virtual health agents
      - AI-enhanced surgical robots
      - Proactive vs reactive care
  - **Impact Areas**
    - Patients
      - Early disease detection
      - Affordable and accessible care
    - Providers
      - Reduced administrative burden
      - Enhanced decision-making
    - Systems and Policy
      - Ethical AI frameworks
      - Data security measures
  - **Predictions for 2030**
    - 1 Million lives saved yearly by AI
    - 1 Billion people using AI for health
    - Advanced telemedicine as the norm
    - CRISPR 3.0 for genetic editing
    - AI-driven predictive care
  - **Predictions for 2040**
    - Fully automated hospitals
    - Neurotechnology in mental health
    - Global pandemic prediction networks
    - AI-enabled medical LLMs
  - **Expanded Capabilities**
    - AI in multimodal diagnostics
    - Dynamic gene editing with AI feedback
    - BCI-driven emotional resilience tools
  - **Ethical Considerations**
    - Equitable access to AI and genetic tools
    - Cultural sensitivity in telemedicine
    - Transparency in AI healthcare applications

PREDICTIONS

- One million lives saved per year due to AI in 2030
- One billion people using AI for health
- Global disease surveillance networks powered by AI
- Personalized nutrition and metabolic optimization
- AI-assisted mental health platforms for global use

WHAT IS IT?

Healthcare will seamlessly integrate advanced technologies across every aspect of patient care, transforming how medicine is practiced and experienced. AI-driven diagnostics will analyze medical data with unparalleled speed and accuracy, identifying patterns and anomalies that might escape even the most experienced clinicians. Wearable devices will continuously monitor vital signs, detect early warning signals for chronic diseases, and provide health feedback to patients and healthcare providers in real-time, fostering a culture of proactive care.

Picture this future: A healthcare system where every person, no matter where they are in the world, has equal access to the finest care. Imagine technology that knows you—not just your symptoms but your history, fears, and hopes. A healthcare journey that is proactive rather than reactive, that is compassionate and precise. This is not the stuff of science fiction; it's a vision built on the concrete advances we see emerging today. The pieces of this future are already being assembled piece by piece, innovation by innovation.

Virtual health assistants will revolutionize patient interactions, handling administrative tasks like appointment scheduling, medication reminders, and initial symptom assessments, allowing doctors to focus on complex decision-making and treatment plans. These AI assistants will also provide personalized health coaching, empowering individuals to take greater control of their well-being.

Genomic medicine will unlock the potential for highly personalized treatments, tailoring therapies to an individual's genetic profile. This will enable healthcare to shift from a reactive model to a predictive and preventive approach, identifying and addressing risks before they manifest into severe conditions. AI-powered systems will also interpret complex medical imaging, such as MRIs and CT scans, enhancing early detection of diseases like cancer or neurological disorders.

These advancements will not only improve accessibility by bridging geographical barriers through telemedicine and remote diagnostics but also elevate the quality of care. Patients in remote or underserved regions will have access to the same expertise as those in urban centers, thanks to virtual consultations and AI-aided diagnostics. By 2030, physical limitations will no longer constrain healthcare, and the fusion of technology with medicine will create a more equitable, efficient, and effective global healthcare system.

WHAT'S GOING TO HAPPEN?

By 2030, healthcare will be redefined by seamlessly integrating cutting-edge technologies, making medicine more precise, proactive, and accessible. AI and AGI systems will be at the heart of this transformation and will act as intelligent assistants to healthcare professionals. These systems will analyze vast amounts of medical data to provide diagnostic insights, recommend personalized treatments, and even monitor patients remotely with exceptional precision and speed. Once prone to delays and human error, diagnostics will be revolutionized by AI-powered algorithms capable of interpreting X-rays, MRIs, and CT scans with superhuman accuracy. Early detection of conditions like cancer and heart disease will become routine, saving countless lives.

The rise of AI-driven wearable devices will empower individuals to take control of their health. These devices, monitoring vital metrics such as heart rate, blood pressure, and glucose levels, will provide real-time alerts for potential health risks before they escalate. Patients will no longer have to wait for annual checkups to catch warning signs; healthcare will become an ongoing and dynamic process. Personalized medicine will reach new heights, as genomic data will enable AI systems to tailor treatments to each patient's genetic profile. This will improve treatment efficacy and significantly reduce side effects, making precision medicine the norm rather than the exception.

In hospitals and operating rooms, AI-enhanced surgical robotics will perform intricate procedures with unmatched precision, minimizing recovery times and risks. AI systems will transform emergency rooms to triage patients based on urgency, predict health declines, and optimize resource allocation, ensuring that critical patients receive timely care. Even mental health, often sidelined in traditional healthcare systems, will receive a much-needed overhaul. AI-powered virtual therapists will provide personalized support and monitoring, making mental healthcare accessible to people around the globe, breaking down stigmas, and addressing the growing mental health crisis.

Digital twins will serve as highly advanced, AI-driven simulations of an individual's body, providing an unparalleled view into their health status and physiology. Picture a digital replica of yourself that integrates data from your wearables, genetic profile, and daily habits—offering insights into how your body would respond to various treatments. This virtual "trial run" eliminates guesswork, allowing physicians to predict the outcomes of interventions like surgeries, medications, or lifestyle changes before they're applied in the real world.

Digital twins will revolutionize personalized medicine by enabling treatment plans that are finely tuned to the individual. These models continuously update based on real-time monitoring from sensors and health devices, allowing adjustments to be made proactively. For example, if a medication is causing side effects or isn't working as intended, the digital twin will identify the issue and suggest refinements, reducing complications and optimizing outcomes. This level of precision will transform healthcare from reactive to highly predictive, ensuring that every patient receives tailored, effective care.

Case Study: Amy's Digital Twin for Personalized Cancer Treatment.

Amy, a 50-year-old diagnosed with early-stage breast cancer, faces many treatment options. Instead of simply guessing which chemotherapy regimen would work best, her oncologist creates a digital twin that simulates her body's reaction to different chemotherapy drugs.

The digital twin evaluates each option based on Amy's genetic predisposition, immune system status, and health metrics. After analyzing multiple simulations, the AI recommends a chemotherapy plan that balances efficacy with minimal side effects. This results in faster remission with fewer complications, enabling Amy to return to her everyday life much sooner.

Telemedicine, turbocharged by AI, will ensure that even the most remote communities can access quality healthcare. Patients will receive preliminary diagnoses, remote consultations, and tailored healthcare plans from the comfort of their homes. With AI-driven virtual reality platforms offering immersive, interactive training for medical students, medical education will also evolve. These tools will simulate complex procedures and real-world scenarios, allowing students to practice and learn without needing physical specimens.

Hospitals will become more efficient with AI-driven robot nurses, which will handle routine tasks such as monitoring vitals, delivering supplies, and assisting in patient care, enabling human nurses to focus on providing emotional support and hands-on care. AI will accelerate drug discovery in the pharmaceutical industry, cutting years off the time required to develop life-saving treatments and making critical medications more affordable and accessible.

By 2030, healthcare will no longer be confined to the walls of a hospital or clinic; it will be an interconnected ecosystem of advanced tools and systems designed to keep individuals healthier for longer. This future will prioritize prevention over reaction, accessibility over exclusivity, and personalization over one-size-fits-all solutions. As these technologies converge, they will create a global healthcare system that is smarter and more humane, addressing the unique needs of every patient with care, empathy, and precision.

WHY DOES IT MATTER?

The advancements shaping healthcare by 2030 have the power to address some of the most pressing challenges humanity faces. At the core of these innovations is a drive to create equitable access to care, breaking down the barriers of geography and inequality. Telemedicine and decentralized tools will allow people in rural or underserved areas to access the same medical expertise as those in major cities. No one will be left behind, as healthcare becomes a global right rather than a privilege.

Efficiency and affordability will be dramatically improved. With AI automating administrative tasks and streamlining diagnostics, healthcare providers can redirect their focus to what truly matters—caring for patients. The cost of care, often a barrier for millions, will decrease as automation and precision reduce waste and eliminate inefficiencies. Families won't have to choose between financial stability and getting the medical help they need.

The result? Improved outcomes. Predictive analytics and personalized medicine will catch diseases early, when they're easier to treat, and offer therapies tailored to individual needs. This isn't just about saving lives—it's about enhancing the quality of those lives, ensuring people can live healthier, longer, and more fulfilling days. Imagine a world where diseases like diabetes or cancer are managed with precision, reducing suffering and offering hope.

As we step into a transformative era of medical innovation, it's crucial to reflect on why these changes matter beyond cutting-edge technology. These advancements are not just about showcasing what machines can do but about reshaping healthcare to center on the human experience. This new frontier is about enhancing lives, bridging gaps, and empowering individuals to live longer, healthier, and more fulfilling lives.

Imagine a world where healthcare is no longer confined to the walls of a hospital or clinic. AI-powered tools, wearable devices, telehealth systems, and digital twins bring care directly to your doorstep. A hospital visit is no longer your first or only line of defence—it's the backup, the safety net. For a single mother in a remote village who previously faced barriers to care or an older adult living in a bustling urban center, healthcare meets them where they are, seamlessly and efficiently. The ability to have real-time monitoring, virtual consultations, and AI-assisted diagnostics at home redefines the patient experience, making healthcare personal and accessible.

Historically, healthcare has been riddled with disparities—those in rural or underserved communities often received care that was worlds apart from what was available in cities. But this new era is about levelling the playing field. With decentralized care, advanced diagnostic tools, specialist consultations, and continuous monitoring become accessible regardless of where you live or your socioeconomic status. For many, this shift represents a convenience and a lifeline, transforming what was once a luxury into a right.

These advancements will unlock global collaboration on an unprecedented scale. AI networks will allow doctors, researchers, and governments to share real-time insights and solutions, enabling faster responses to global health crises like pandemics. Shared medical knowledge will no longer be siloed but rather a tool for collective progress, benefiting every corner of the world.

Ultimately, these innovations matter because they promise a future where healthcare is more intelligent, compassionate, and accessible. It's about building a system that not only cures but cares, bringing us closer to a world where no one is left behind.

WHO WILL IT IMPACT?

The future of healthcare will touch every aspect of society, starting with the people who matter most—patients. With AI-driven diagnostics, wearable devices, and personalized treatment plans tailored to genetic profiles, healthcare will shift from reactive to proactive. Diseases will be caught earlier, chronic conditions will be better managed, and care will be made more affordable. Imagine a world where waiting for test results is obsolete, and access to lifesaving treatments is universal, not dictated by geography or income.

For healthcare providers, the transformation offers relief from administrative burdens. AI will handle scheduling, records, and triaging, freeing up doctors and nurses to focus on what truly matters—caring for people. Real-time data will enable more intelligent decisions, improving outcomes and easing the burnout that has plagued the profession for decades. This shift will let providers return to the essence of medicine: human connection.

Healthcare systems will face new responsibilities to ensure ethical AI use and robust data security. Patients must trust that their information is safe and that AI systems work somewhat and without bias. Governments will need to craft policies addressing data ownership and equitable access questions, while innovators must prioritize inclusivity and transparency as they redefine what's possible in medical care.

From patients to providers, policymakers to tech companies, this transformation is a collective opportunity to build a healthcare system that is faster, fairer, and more human. Technology won't replace compassion—it will amplify it, creating a future where care is accessible, personal, and driven by the shared goal of healthier, longer lives for all.

CALL TO ACTION

It's time for governments, businesses, and medical institutions to act decisively to prepare for the AI-driven healthcare revolution. Start by creating specific frameworks to govern how AI is used in medical settings—rules prioritizing patient data privacy, ensuring unbiased algorithms, and protecting against misuse. These frameworks shouldn't just exist on paper; they must be actionable and enforced, providing clarity for innovators and accountability for everyone involved.

Invest heavily in AI-focused R&D—not just in the shiny, headline-grabbing applications but in the infrastructure and foundational technologies that make them possible, like secure data systems and advanced training platforms for medical professionals. Hospitals and medical schools should immediately begin implementing AI training programs, ensuring doctors, nurses, and administrators know how to leverage these tools in patient care effectively.

At the same time, governments must create funding incentives for deploying AI in underserved areas. AI tools can reduce healthcare gaps only if they're made accessible to rural clinics, overburdened urban hospitals, and low-income communities. Without this focus, the technology will widen inequalities instead of bridging them.

Finally, businesses developing healthcare AI must prioritize transparency—publish how their algorithms work, provide clear data security measures, and involve diverse voices in the design process to ensure these systems work for everyone. This is not just about innovation; it's about trust, safety, and the future of how we care for one another.

The future of healthcare is not inevitable; it's created by our actions today. Whether you're a **healthcare professional**, a **patient**, a **policymaker**, or an **innovator**, there are steps you can take to be part of this transformative journey:

- **Embrace the Tools:** Use the wearables, the apps, and the virtual care options available today. Get comfortable with these technologies because they are the stepping stones to an empowered future.

- **Demand Ethical Standards:** Advocate for transparency, privacy, and fairness in AI-driven healthcare. Ensuring these systems work for all people equitably is key to realizing their full potential.

- **Collaborate and Share Knowledge:** Innovators and researchers, continue sharing your insights, discoveries, and tools. Collaboration across borders, disciplines, and sectors will be the key to making healthcare a unified human achievement.

The future of healthcare is hopeful. It is bright with possibilities, powered by the brilliant convergence of human compassion and technological advancement. Imagine a world where healthcare is not something you fear but something that quietly supports your life, always there in the background, guiding you, keeping you safe, and helping you live your fullest life.

A world where being healthy is not about privilege but **human dignity**—where a child born in a remote village has the same chance at a healthy life as a child born in a bustling metropolis—a world where healthcare systems are connected, empathetic, and proactive.

This is the world we are building. Our vision drives innovation today, the dream that motivates doctors, researchers, patients, and caregivers alike. We are creating a healthcare system that doesn't just heal but inspires, ensuring every person can lead a healthier, happier life.

The future is coming fast, bringing opportunities we cannot fully grasp. But one thing is sure: this future belongs to all of us, and together, we can make it a reality.

LET'S EMBRACE IT. LET'S BE PART OF IT.

A New Industrial Revolution in Healthcare

The transformation unfolding in healthcare today mirrors the seismic shifts of the Industrial Revolution—but instead of factories and machines, it's AI and digital technologies reshaping our world. Much like the Industrial Revolution brought mass mechanization and scalable production, AI is driving a profound shift from one-size-fits-all medicine to deeply personalized care. Treatments are no longer generic solutions but tailored to an individual's unique genetic, physiological, and behavioural profile. This transition marks a leap forward in precision and humanity in medicine.

During the Industrial Revolution, innovations like steam engines and railroads bridged vast geographic divides, connecting remote regions to urban centers and enabling unprecedented access to goods and services. Today, AI, telemedicine, and edge computing are playing a similar role in healthcare, bridging gaps in access by linking underserved areas to advanced medical knowledge and resources. A rural clinic with limited specialists can now tap into global expertise, making quality healthcare accessible no matter where you are.

AI: Both a Disruptor and a Unifier

AI is both a disruptor and a unifier in the evolving healthcare landscape. As a disruptor, AI challenges traditional roles within medicine. Tasks once performed solely by human physicians—like analyzing medical images or interpreting complex datasets—can now be enhanced or even led by AI systems, often faster and more accurately. It doesn't replace doctors but augments their capabilities, enabling them to focus on patient care while AI handles the data-heavy lifting.

At the same time, AI serves as a unifier. It connects physicians and researchers across the globe, creating shared pools of medical insights. Consider rare diseases that may have only a handful of cases worldwide—AI can consolidate this scattered information into coherent datasets, giving specialists a more transparent, more comprehensive understanding of these conditions. This global collaboration, powered by AI, turns fragmented knowledge into actionable insights.

The Transformative Patient Experience in 2030

Picture a healthcare experience entirely reimagined by technology: You enter a clinic, and without uttering a word, the system already knows your medical history, predicted risks, and current treatments. Digital receptionists guide you seamlessly through the process while AI crafts a treatment plan tailored to your exact needs. Visits that once involved tedious administrative tasks and impersonal care now feel efficient, intuitive, and deeply personalized.

Case Study: Healthcare Without the Travel

Take Maria, a 45-year-old living in a remote village. In 2023, she had to travel over 50 miles for a basic health consultation—a journey that consumed her time and resources. By 2030, AI-powered diagnostics and teleconsultations have transformed her life. Maria connects with physicians from the comfort of her home through AI-enhanced platforms. Her wearable device continuously updates her digital health profile, allowing her doctor to make informed decisions in real-time.

This new era in healthcare isn't just about technology—it's about bridging divides, empowering individuals, and creating a system that prioritizes humanity and precision. Like the Industrial Revolution, this transformation will redefine how we live, how we work, and, most importantly, how we care for one another.

Case Study: From Reactive to Predictive Care

John, a 60-year-old man with a history of hypertension, used to experience healthcare reactively—visiting his doctor only when symptoms worsened, or emergencies arose. But in 2030, John's care will be transformed by an AI-driven health platform that turns episodic interactions into a continuous, predictive journey toward wellness.

At the heart of John's transformation is a digital twin—a virtual replica of his body created from data gathered by his wearable health sensors. This digital twin monitors his vitals, analyzing diet, activity, sleep patterns, and stress markers. One day, the system identifies subtle signs that John's hypertension might worsen, flagging trends that even a trained physician might overlook during a routine visit.

The AI assistant doesn't wait for problems to escalate. It immediately sends John actionable suggestions, such as adjusting his diet, adding light physical activity to his daily routine, and practicing mindfulness exercises to lower stress. Simultaneously, it notifies his physician, who uses the AI-generated insights to fine-tune John's medication regimen remotely. The proactive intervention prevents what could have been a significant health crisis, such as a stroke or heart attack.

With AI constantly monitoring his health and providing timely insights, John's healthcare experience shifts from reactive to predictive. He avoids hospital visits and the complications that come with unmanaged hypertension. His quality of life improves as he feels empowered to make healthier choices guided by real-time data. For John, 2030 isn't just about advanced technology—it's about living a longer, healthier life thanks to a healthcare system designed to keep him well, not just treat him when he's unwell.

This new paradigm of predictive care underscores how AI will transform healthcare—not only for John but millions of people worldwide, making prevention the cornerstone of medical practice.

Key Values from Technology in John's Predictive Care Journey:

- **Proactive Prevention.** Technology enables early detection of subtle health trends, turning reactive emergency care into proactive prevention. This shift allows interventions like dietary adjustments and medication tweaks before conditions escalate, reducing the risk of severe health crises such as strokes or heart attacks.

- **Personalized Healthcare.** Digital twins and AI-driven insights tailor healthcare to John's unique needs, ensuring treatments and lifestyle recommendations are optimized for his physiology and daily habits. This personalization enhances the effectiveness of care while minimizing unnecessary interventions.

- **Empowerment Through Data.** Real-time monitoring and actionable insights empower John to take control of his health. With constant guidance from AI and collaboration with his physician, he can make informed decisions, improving his quality of life and fostering a sense of agency over his well-being.

Virtual AI Agents

Virtual agents will transcend their origins as basic chatbots, evolving into empathetic, adaptive companions capable of deeply understanding a person's emotions, motivations, and needs. These AI systems will grow alongside patients, learning their preferences and health journeys to provide personalized support. Imagine a virtual agent that not only answers your medical queries but also knows how to speak to you during moments of anxiety or doubt, providing comfort and guidance tailored to your personality and circumstances.

Intelligent virtual agents will be pivotal in mental health care and lifestyle management. They'll do more than send reminders or track vitals—they'll actively engage in meaningful conversations that promote emotional well-being and adherence to treatment plans. For example, using voice analysis, an agent could detect signs of stress or fatigue, suggest relaxation techniques and breathing exercises, or even schedule a virtual therapy session. When a patient struggles with lifestyle changes, the agent can offer motivational insights, track progress, and celebrate milestones, fostering a sense of accomplishment and support.

In 2030, intelligent virtual agents will redefine the patient experience, merging cutting-edge AI with human-like interaction to make healthcare more compassionate, proactive, and effective.

Hypothetical Case Study: Emma's Battle with Anxiety – Assisted by a Virtual Health Agent

Emma, a 30-year-old struggling with anxiety, finds it challenging to keep up with therapy sessions. Her AI-driven health agent is not just a reminder—it provides guided breathing exercises when Emma's wearable detects signs of stress. It also simulates conversational therapy based on her preferences, offering empathetic, interactive support during difficult times. When Emma hesitates about attending a therapy session, the agent nudges her positively by discussing the benefits she gained from previous sessions. Over time, Emma finds herself more engaged in her treatment, leading to fewer anxiety attacks and an overall improvement in her mental health.

KEY VALUES:

- **Empathy in AI:** Virtual health agents bring a layer of human-like understanding to healthcare, providing constant and personalized support.

- **Behavioural Insights:** These agents collect and analyze behaviour patterns to anticipate needs, promoting a proactive approach to managing health.

PREDICTIONS FOR 2030

1. **AI-Driven Diagnostics**
   AI will integrate multimodal data—imaging, genetic profiles, and real-time health metrics from wearables—to provide early, precise diagnoses. Predictive health scores will allow patients and doctors to address risks like cardiovascular diseases long before symptoms appear. This fusion of data will revolutionize disease prevention and treatment.

2. **Telemedicine as the Default**
   Virtual care will become the standard for initial healthcare interactions, leveraging AI-driven triage tools and global specialist networks. Patients will receive diagnoses, consultations, and treatments from anywhere, enhancing access and efficiency, especially for underserved areas.

3. **CRISPR Advancements**
   CRISPR 3.0 will enable reversible gene editing, adding safety and adaptability to genetic therapies. Real-time AI monitoring will optimize

treatments for conditions like cystic fibrosis or sickle cell anemia, dynamically adjusting edits to meet evolving patient needs.

4. **Wearable IoT Revolution**

Advanced wearables will act as proactive "health guardians," monitoring physiological data and providing real-time alerts for early intervention. These devices will connect seamlessly with healthcare providers, offering tailored advice and preventing severe health crises like strokes or heart attacks.

5. **Personalized and Predictive Medicine**

Healthcare will move toward hyper-personalized treatment plans powered by AI and genomic tools. Predictive analytics will forecast health risks, enabling preventive care strategies tailored to individual needs reshaping medicine from reactive to proactive.

PREDICTIONS FOR 2040:

1. **Fully Automated Healthcare Facilities**

By 2040, hospitals will operate like precision machines, leveraging AI and robotics for nearly every aspect of patient care. From AI-powered diagnostic systems to robotic surgeries enhanced with augmented reality, healthcare facilities will provide 24/7 care without fatigue or error. These advancements ensure unmatched precision in treatments, faster recovery times, and consistent care, regardless of time or place.

2. **LLM-Enabled Physicians**

Large language models will revolutionize patient-physician interactions by providing instant, evidence-based recommendations and breaking complex medical information into understandable language. These AI partners will enhance empathy, ensuring patients feel heard and understood while equipping doctors with the latest insights for personalized care.

3. **Neurotechnology in Mental Health**

Brain-computer interfaces will transform mental health care, offering real-time monitoring of emotional states and precise interventions for conditions like PTSD and anxiety. These technologies will empower patients to regain control over their mental health through personalized therapies and immediate support, fostering a sense of autonomy and hope.

4. **Global Pandemic Prediction Networks**

AI-driven surveillance systems and IoT health sensors will form a global safety net, identifying emerging pathogens before outbreaks escalate. By analyzing real-time data and coordinating international responses, these networks will ensure swift containment of pandemics, safeguarding lives and restoring trust in global health systems.

EXPANSION AND DEEPER POINTS:

- **AI's Role in Multimodal Diagnostics:** AI's ability to synthesize multiple data streams is akin to connecting the dots across different diagnostic domains. For instance, AI can integrate a patient's imaging data (like MRIs), lab test results, genomic data, and even real-time monitoring from wearables. This holistic approach to diagnostics provides insights that would be impossible to obtain from any single data source. AI identifies diseases earlier and distinguishes subtle variations between similar conditions—ensuring tailored treatments.

- **Beyond Diagnosis—Predictive Analytics for Preventive Care:** AI in diagnostics is not merely about diagnosing existing conditions—it's about predicting future health risks. Predictive models use historical data to identify trends and warn of emerging health problems before symptoms appear. For instance, AI can detect early warning signs of Alzheimer's by analyzing language patterns from voice data or subtle changes in gait, providing opportunities for early intervention that were never possible before.

- **Reducing Diagnostic Disparities:** Another profound impact of AI-driven diagnostics is its role in lowering diagnostic disparities. Many rare diseases go undiagnosed or misdiagnosed simply because healthcare professionals may have limited exposure to them. AI's training on vast datasets, including rare conditions, helps overcome this barrier, giving equitable access to precise diagnostics across communities, regardless of a physician's experience level.

- **Adapting Telemedicine for Cultural Sensitivity:** A significant challenge and opportunity for telemedicine lies in cultural adaptation. Different regions have unique attitudes toward healthcare, privacy, and communication. AI-powered translation and cultural adaptation algorithms can adapt consultations to match local norms—ensuring that healthcare is available and accessible in culturally appropriate ways. A telehealth session in Japan might include more formal speech, whereas, in other regions, a more casual conversation could foster trust.

- **Overcoming Digital Barriers:** While telemedicine presents an incredible opportunity, digital infrastructure gaps remain, particularly in developing countries. Innovators are working on lightweight telehealth apps on 2G networks or via SMS for areas without robust internet. These solutions promise telemedicine to even the most digitally disconnected places, ensuring no one is left behind.

- **Virtual Reality Consultations for Complex Cases:** By 2040, telemedicine will include VR consultations that provide a more immersive experience, helping physicians conduct more nuanced remote examinations. Patients with mobility issues will benefit from virtual rooms where they can interact with healthcare professionals more dynamically, allowing for better assessments of things like posture, gait, and physical behaviour.

- **AI-Driven Early Warnings for Chronic Diseases:** Wearables in 2040 will not just track vitals—they'll anticipate chronic conditions like Type 2 diabetes or hypertension by analyzing years of data. Continuous monitoring allows AI to detect slow-developing problems that annual checkups might miss, providing lifestyle recommendations to prevent diseases entirely.

- **Social Integration and Behavioral Nudges:** Wearables will also integrate with social platforms and provide positive behavioural nudges based on the wearer's social preferences. For example, a wearable could notify someone to join a friend's exercise session or take a group walk—promoting mental and physical health through social integration.

- **Mental Health Monitoring:** These devices won't only focus on physical metrics but also on mental well-being. Sensors could track skin conductivity, sleep patterns, and heart rate variability to gauge stress levels. The wearable could prompt users to take breaks or initiate a guided meditation when stress markers react as mental health guardians in daily life.

- **Ethics and Equity in Genetic Editing:** While CRISPR offers profound possibilities, ethical challenges exist. Who gets access to gene editing? Will it be something only the wealthy can afford, leading to more significant health disparities? These are not just technical issues—they are societal ones. Ensuring equitable access to CRISPR technology and preventing a divide between those who can and cannot afford genetic modifications must be addressed through policy and global cooperation.

- **CRISPR and Personalized Medicine:** CRISPR won't be limited to treating monogenic diseases like cystic fibrosis. By 2040, the technology will tackle more complex, multi-gene conditions like autism or diabetes. AI will help analyze an individual's genetic profile and determine which edits, if any, could mitigate risk factors for these complex diseases, ushering in a new era of personalized medicine that is not just reactive but preemptive.

- **Dynamic Gene Editing with AI Feedback:** The combination of CRISPR and AI will allow for dynamic gene editing, where gene editing is monitored and adjusted based on real-time patient condition feedback. Treatment can be refined continuously instead of a one-time intervention, offering the best possible outcomes and responding to unforeseen genetic interactions.

- **Direct Neurofeedback for Cognitive Enhancement:** BCIs will provide real-time, data-driven neurofeedback that helps patients effectively identify and work through emotional triggers. For instance, patients with chronic depression can use BCIs to visualize their brain's response to different positive stimuli, effectively rewiring their neural pathways to adopt healthier emotional patterns.

- **BCI-Driven Emotional Resilience Training:** By 2040, BCIs won't just be used in therapy sessions; they'll become a part of everyday wellness. People will use BCIs in their homes to practice emotional regulation—using guided neurofeedback to build emotional resilience. This could be particularly effective for frontline workers or those experiencing prolonged stress, allowing them to train their brains to recover from stressors more quickly.

- **Therapist-AI Collaboration:** Therapists will collaborate with AI-powered BCIs to better understand their patients. The AI component can analyze brain data and identify cognitive patterns correlating negative thought processes. This will allow therapists to target their interventions precisely where patients need them most, enhancing therapeutic efficacy.

Energy and Climate

In the vibrant, sustainable world of 2030, Brad and Caitlin represent two seemingly opposing forces converging for a shared purpose: a cleaner, healthier planet. Brad, a 27-year-old climate activist, spends his nights decoding the vulnerabilities of oil and gas corporations. His mission is to highlight energy systems that cause environmental degradation

Caitlin, on the other hand, is a seasoned investor in clean energy. At 45, she's spent decades championing nuclear power, advocating for its role as a misunderstood hero in the fight against climate change. She focuses on Small Modular Reactors (SMRs), which she sees as humanity's best chance to provide abundant, low-carbon energy to billions while phasing out fossil fuels. Caitlin works tirelessly to shift public perception, often hosting workshops and leading campaigns to demystify nuclear energy's potential and safety.

Brad and Caitlin met at a global energy summit in Geneva, where activists, investors, and innovators convene to discuss the next wave of sustainable solutions. Brad, there to challenge corporate inertia, listens skeptically as Caitlin delivers a passionate speech about SMRs. She describes their ability to provide clean energy to underserved regions and their potential to stabilize grids overloaded by intermittent renewables. Initially dismissive of nuclear energy, Brad finds himself intrigued by Caitlin's data-driven arguments and her acknowledgment of the energy industry's need for transparency.

After her talk, Brad approaches Caitlin, starting with a sharp critique of corporate greed in the energy sector. Caitlin counters calmly, agreeing with his frustrations but emphasizing the need for pragmatic, scalable solutions. Their heated exchange evolves into a collaborative discussion, with Brad explaining how AI could enhance SMR efficiency.

Their conversations lead to a groundbreaking partnership. Brad, leveraging his expertise in AI and cybersecurity, begins working with Caitlin's team to optimize SMR systems. He develops algorithms to identify potential vulnerabilities in reactor networks, ensuring they are resilient against cyberattacks. Caitlin mentors Brad in navigating institutional frameworks, showing him how to channel his activist fervour into actionable policy change.

Together, they create a platform that merges Caitlin's investment in nuclear energy with Brad's AI innovations. The platform tracks real-time energy output and carbon savings from SMRs while integrating Brad's transparency tools to allow public oversight of energy production. This transparency shifts public opinion, fostering trust in nuclear energy as a viable, safe alternative.

Their collaboration doesn't just impact energy systems; it inspires a global movement. Once skeptical of atomic energy and tech-driven activism, communities begin to embrace these solutions' possibilities.

By the end of 2030, Brad and Caitlin's partnership has contributed to a 15% reduction in global carbon emissions. Their shared vision—a world powered by clean, accessible energy without compromising ethics or humanity—is a beacon for what's possible when passion meets pragmatism.

In this new era, Brad and Caitlin prove that even the most divergent paths can converge to create transformative change, unlocking humanity's potential while safeguarding the planet for future generations.

ENERGY & CLIMATE IN 2030

 # Energy and Climate 2030

- **Introduction**
  - Cost of energy
  - Role of scarcity and demand
  - Sustainable energy as a key to abundance

- **Key Players**
  - Brad: Climate activist and hacker
    - Disrupts fossil fuel systems
    - Advocates ethical hacking and AI
  - Caitlin: Clean energy investor
    - Focus on Small Modular Reactors (SMRs)
    - Advocates for nuclear energy transparency

- **Predictions for 2030**
  - Continued CO2 emissions growth
  - Dominance of coal in developing countries
  - Expanded use of AI-optimized nuclear energy
  - Arrival of AGI and ASI for energy
    - AGI: Smart grids
    - ASI: Near-zero-cost energy

- **Energy and Power Overview**
  - Definitions
    - Energy as work
    - Power as rate of work
  - Sustainable energy goals
    - Minimize CO2 emissions
    - Orders of magnitude in energy use

- **Challenges and Opportunities**
  - Rising global energy demand
  - Scaling clean energy technologies
    - Nuclear (SMRs)
    - Solar and wind
  - Addressing energy poverty
    - Uneven global energy access

- **Innovations and Solutions**
  - AI-driven energy systems
    - Grid optimization
    - Predictive energy management
  - Clean energy technologies
    - SMRs for decentralized grids
    - Hydrogen from high-temperature reactors
  - Role of digital economy
    - Reducing reliance on physical goods
    - Virtualization of services

- **Global Energy Transition**
  - Current trends
    - Increased electricity demand
    - Solar and wind growth
  - Role of governments
    - Policies for renewables and nuclear
    - Addressing geopolitical energy challenges

- **Future Outlook**
  - Path to energy abundance
    - Clean, reliable, low-cost energy
    - Equitable energy access
  - Importance of collaboration
    - Governments, industries, and individuals
    - Balancing innovation and inclusivity
  - Call to Action
    - Advocacy for sustainable policies
    - Investment in green technologies

PREDICTIONS FOR 2030

- The world will not meet the UN Access to electricity – SDG7: Energy goal to ensure that everyone has access to affordable, reliable, sustainable, and modern energy.

- Data centers, driven by AI, will consume 5-9% of U.S. electricity generation.

- CO_2 emissions will continue to grow in 2030. There has been a continued demand for sustainable energy solutions, but coal will be a dominant energy source - especially for developing countries.

- Humanity's energy use will continue to grow – especially in non-OECD countries.

- Expanded use of nuclear energy optimized by AI for lower costs and better efficiency.

- Other forms of sustainable energy will begin to emerge: fusion, hydrogen, renewable storage, antimatter, solar beaming, etc.

WHAT IS IT

Energy is the Ability to Do Work, measured in many different units—one reason why it is so complicated: Joule (J), calorie (cal), watt-hour (Wh), British thermal unit (BTU), barrel of oil equivalent (boe), natural gas cubic foot (cf).

Power is the Rate Work is Done (E/T). It is measured in Watts (w).

Sustainable energy that minimizes CO_2 emissions is the key to humanity's abundance.

In Infinite Progress: *How the Internet and Technology Will End Ignorance, Disease, Poverty, Hunger, and War,* Byron Reese wrote "the cost of making almost everything is mostly energy and intellect, not raw materials. The energy—sometimes human exertion, sometimes mechanical exertion—has always come at a cost. But what if that energy cost fell to zero? As my economics professors insisted, cost is determined by scarcity and demand. But is energy really scarce—or is it like air? Is it finite, or is it for all practical purposes infinite?"

150 ZJ of energy are available to humanity from the sun, approximately 10,000 times more than we used in 2023. What abundance could we have if we only learned to harness this energy free from the sun?

In his chart, "Electricity Use Defines The World", Robert Bryce outlines three different energy worlds that humans live in:

1. **Global Population Breakdown (7.9 billion total):**
   - Unplugged World (47%): 3.7 billion people consume ≤1,200 kWh/capita/year, roughly the energy needed to power a large U.S. kitchen refrigerator annually.

   - Low-Watt World (16%): 1.3 billion people use between 1,200 and 4,000 kWh/capita/year.

   - High-Watt World (37%): 2.9 billion people use >4,000 kWh/capita/year.

2. **Global Average Electricity Consumption: Approximately 3,500 kWh/capita/year.**

Per Bryce in *Burned Alive,* a large U.S. refrigerator consumes around 1,200 kWh annually, the same as the "Unplugged World" total annual consumption. This underscores the disparity in electricity access and usage globally, with a significant portion of the population in the "Unplugged World" category.

As of the third quarter of 2024, the global average electricity prices were approximately $0.18 per kilowatt-hour (kWh) for households and $0.15 per kWh for businesses (https://www.globalpetrolprices.com/electricity_prices/).

Countries with Low Electricity Prices:

- **Iran:** ~$0.002 per kWh
- **Qatar:** ~$0.03 per kWh.
- **Saudi Arabia:** ~$0.05 per kWh.
- **China/India:** ~$0.076 per kWh

Countries with Highest Electricity Prices:

- **Ireland:** ~$0.42 per kWh.
- **Italy:** ~$0.44 per kWh.

These variations are influenced by factors such as energy production methods, government subsidies, taxes, and each country's overall energy policy framework.

WHY IT MATTERS

The availability of low-cost energy will determine whether there will be a world of abundance. Global climate change threatens ecosystems, economies, and human health. Energy transition and efficiency are keys to reducing carbon emissions, ensuring sustainable development, and mitigating climate risks.

Rising energy demand is driven by building heating and cooling, electric vehicles, AI/data centers, and the energy poor of the world wanting abundant energy today had by the energy-rich will be one of the most significant challenges for humanity in the next five to 25 years. If we can innovate and solve for energy, most other challenges will be solved, and we can have a world of abundance; if we cannot solve the sustainable energy problem, most other issues will not be solved.

WHO IT WILL IMPACT

Governments will be at the forefront, tasked with reimagining energy infrastructure and implementing policies to combat climate change. They must navigate the complex intersection of energy security, sustainability, and economic growth. Policies fostering renewable energy adoption, funding for nuclear innovation, and incentives for clean technology startups will be critical. Governments must also address the geopolitical implications of energy transitions, such as reducing dependence on fossil fuel exports and securing the rare materials essential for clean energy technologies. Those who embrace these changes early will lead in global climate diplomacy, while laggards may face increasing economic and environmental pressures.

Energy-intensive industries, particularly in AI, data centers, and energy production, will see significant impacts as sustainability becomes a competitive advantage. Companies developing advanced AI models will rely on greener data centers powered by renewables or next-generation nuclear energy to reduce their carbon footprint. Energy corporations must adapt to decentralized grids, integrating next-generation nuclear and small modular reactors (SMRs), solar power, and energy storage technologies. This era will offer unprecedented opportunities for tech innovators to lead in areas like AI-driven energy optimization, innovative grid technologies, and carbon capture systems.

Citizens will experience these changes most tangibly in their daily lives. Energy costs could stabilize or decrease as efficient, decentralized systems replace traditional grids, providing affordable access to clean energy. Urban and rural communities will benefit from decentralized energy solutions, reducing outages and improving reliability. Sustainability efforts will shape consumer choices, from electric vehicles powered by clean grids to energy-efficient smart homes. Moreover, the integration of transparent, AI-powered energy monitoring tools will empower individuals to participate actively in reducing their carbon footprint.

CALL TO ACTION

To prepare for the challenges and opportunities of an increasingly energy-intensive world, it is crucial to take proactive steps today. Governments, businesses, and individuals must champion the adoption of energy-efficient technologies and sustainable solutions. This includes scaling up investments in renewable energy systems, like solar and wind while embracing innovative options such as nuclear power—including next-generation nuclear and small modular reactors (SMRs)—as a reliable, clean, and scalable energy source. During this transition, natural gas will play an important role to keep the price of energy low.

By advocating for policies that support green energy infrastructure, fostering public-private partnerships, and promoting consumer awareness about sustainable energy options, we can collectively build a future where energy is abundant, affordable, and environmentally friendly. Embracing these technologies now ensures a secure energy supply and a meaningful step toward mitigating climate change and protecting the planet for future generations.

DOUG HOHULIN ON ENERGY AND CLIMATE

The Cellular Industry's Role in Creating Abundance and Reducing Emissions and AI's Opportunity

While working at Motorola and Nokia, I had the opportunity to participate in various smart grid and energy projects. The telecommunications industry has driven remarkable progress in economic development and connectivity and plays a pivotal role in creating sustainable solutions for a world striving for abundance. While the cellular industry, like all sectors, generates carbon emissions, it paradoxically enables massive reductions in emissions across other industries. As AI becomes increasingly embedded in energy and communication networks, it has the potential to further accelerate these positive impacts throughout the 2020s and 2030s.

Mobile technology was a key economic development and sustainability enabler in the 2000s and 2010s. According to GSMA Intelligence, a 10% increase in mobile subscribers in developing countries correlates with a GDP increase of 0.8% to 1.2%.

Access to mobile broadband drives new opportunities, especially in remote work, telemedicine, and virtual education. For regions with limited access to physical infrastructure, mobile networks offer scalable, energy-efficient solutions. This is critical as economic growth in the digital domain requires a different level of resource consumption than traditional physical development, meaning the world can grow without a proportional increase in emissions.

A profound example of this shift is the ability to replace physical travel with telepresence. Large conferences and business travel contribute significantly to global CO_2 emissions. For instance, a person attending a three-day conference can generate one ton of CO_2, while the average U.S. annual per capita emissions are 13.7 tons. In contrast, a one-hour Zoom meeting with two participants produces only 3.7 grams of CO_2, equivalent to driving just 52 feet. The cost of a 5,000-person physical conference exceeds $1 million and results in 100-200 tons of CO_2 emissions. Telepresence thus emerges as one of the greenest forms of travel: the safest kilometre or mile is the one travelled virtually. (https://educatorsinvr.com/2020/03/09/green-conference-reducing-carbon-emissions-with-a-virtual-conference/)

The cellular industry has already demonstrated how enabling technologies reduce emissions by nearly a factor of ten. According to Telefónica's research, mobile technologies facilitated a reduction of 2,135 million tons of CO_2 in 2018, almost the same as the total annual emissions of Russia. This means that while mobile networks do have a carbon footprint, their contribution to reduce emissions by nine times that far outweighs the impact that they generate. Mobile networks reduce energy consumption by facilitating remote work, optimizing transportation, and streamlining industrial processes through machine-to-machine (M2M) communications and IoT. Intelligent energy systems powered by mobile and AI technologies allow fuel savings of over 521 billion litres globally, and 1.44 billion MWh of energy have been conserved.

ENERGY AND CLIMATE 2030

The U.S. Information Administration's (EIA) International Energy Outlook 2023 outlines:

"The global energy system is governed by complex dynamics that play out over time across regions and sectors of the economy. Projected increases in population and incomes drive our expectation of rising energy demand through 2050. However, we expect the increased energy demand to be moderated by reduced energy intensity: less energy will be required for each unit of economic activity. In addition, we expect reduced carbon intensity—largely driven by the wide-scale deployment of renewables for electricity generation—which will help limit global CO_2 emissions associated with record-high energy demand. Our International Energy Outlook 2023 (IEO2023) explains our findings and showcases key regional and sectoral variations. We use EIA's detailed World Energy Projection System to produce IEO2023, giving our readers a unique view into future global energy systems." The challenge we face as modellers is delivering actionable insights about the future in a world of uncertainty."

This is highlighted in Figure 3 of the report, which shows that GDP, Primary Energy Use and CO_2 emissions will continue to rise into the 2030s to 2050s if nothing changes.

FIGURE 3

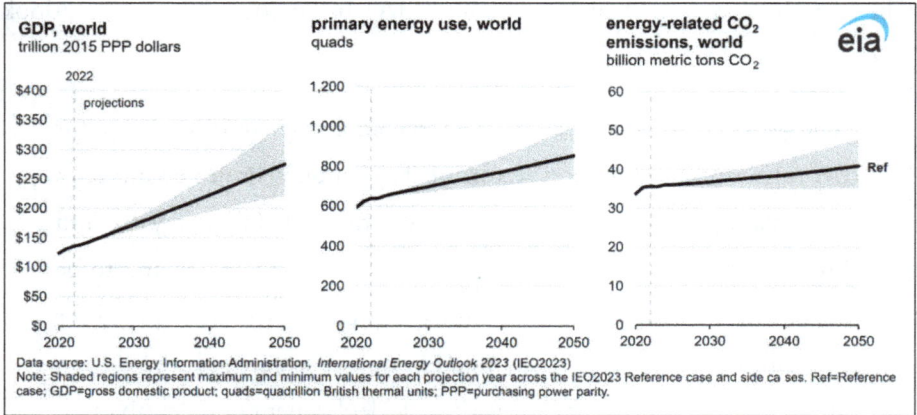

GDP, world
trillion 2015 PPP dollars

primary energy use, world
quads

energy-related CO_2 emissions, world
billion metric tons CO_2

Data source: U.S. Energy Information Administration, *International Energy Outlook 2023* (IEO2023)
Note: Shaded regions represent maximum and minimum values for each projection year across the IEO2023 Reference case and side ca ses. Ref=Reference case; GDP=gross domestic product; quads=quadrillion British thermal units; PPP=purchasing power parity.

INTERNATIONAL ENERGY OUTLOOK 2023. TABLE A4: BILLION OF DOLLARS, 2015. WORLD GDP BY REGION EXPRESSED IN MARKET EXCHANGE RATES

Region	2022	2025	2030	2035	2040	2045	2050
Americas	$28,078	$29,194	$32,220	$35,342	$38,990	$43,033	$47,427
United States	$20,671	$21,361	$23,436	$25,653	$28,414	$31,534	$34,962
Other Americas	$2,516	$2,708	$3,090	$3,496	$3,945	$4,439	$4,976
Europe and Eurasia	$22,949	$23,897	$25,663	$27,295	$29,103	$31,024	$33,091
Asia Pacific	$32,233	$36,392	$44,057	$51,533	$58,587	$65,880	$72,554
Africa and Middle East	$5,526	$6,049	$6,890	$7,795	$8,692	$9,570	$10,432
World	$88,786	$95,533	$108,829	$121,965	$135,372	$149,508	$163,503

For AI, there is an excellent opportunity for economic growth, as outlined in the article "The Promise of Artificial Intelligence – Center for Data Innovation" by Daniel Castro and Joshua New. (https://datainnovation.org/2016/10/the-promise-of-artificial-intelligence) In 12 countries surveyed, AI would boost labour productivity rates by 11% to 37%.

The McKinsey Global Institute estimates the potential economic impact of automating knowledge work to be in the range of $5.2 trillion to $6.77 trillion; advanced robotics, $1.7 trillion to $4.5 trillion; and autonomous/semi-autonomous vehicles, $0.2 trillion to $1.9 trillion.

According to Accenture, by 2035, AI could increase the annual growth of U.S. and Finnish economies by 2% points; the Japanese economy by 1.9 points; and the German economy by 1.6 points.

AI offers even more significant potential to accelerate emissions reductions as we look to the future. AI can further optimize energy grids, predict demand patterns, and minimize waste, creating more efficient systems across industries. As Nokia's Nishant Batra emphasized, network traffic may grow 27 times between 2019 and 2030, yet the telecommunications industry is committed to reducing greenhouse gases by 50% within the same period. By using AI to manage infrastructure efficiently, industries can decouple growth from emissions—scaling digital infrastructure by a factor of 10 while reducing emissions from physical activities. (https://www.nokia.com/blog/network-evolution-towards-the-6g-era/)

By 2040, the integration of AI, IoT, and mobile technologies will power a world where data centers, connected vehicles, and smart cities become integral to sustainable development. This shift toward virtual solutions and AI-optimized operations provides a roadmap to achieving abundance for all—clean, affordable energy that becomes "too cheap to meter." with reduced emissions and improved quality of life. We can unlock an era of sustainable prosperity through strategic investment in AI and cellular networks, minimizing environmental impact while maximizing human potential.

SCALING THE DIGITAL ECONOMY TO REDUCE CARBON EMISSIONS AND CREATE ABUNDANCE

The path to a sustainable future lies in prioritizing the growth of the digital economy over the physical economy. Historically, growth in the physical economy—through manufacturing, infrastructure, and transportation—has been tied to rising carbon emissions. If we continue to grow the physical economy by 2X, we will inevitably see an increase in CO_2 emissions, exacerbating climate change. However, the digital economy offers a promising alternative. If we scale the digital economy by a factor of 10X, we can reduce the size of the physical economy, cut carbon emissions, and still provide abundance for all. The key lies in shifting from physical consumption to virtual experiences, services, and digital infrastructure.

Consider the concept of "virtualization" to minimize our environmental footprint. Today, cloud computing, AI, and telepresence tools allow us to substitute many physical activities—like commuting, conferences, and in-person meetings—with virtual alternatives. The greenest and safest mile you can travel is the one you travel virtually. This reduces the need for transportation and real estate and cuts emissions by orders of magnitude.

By 2040, the world could have 20 trillion objects. If these objects remain physical, their production and disposal will burden the environment. However, we can drastically reduce material consumption by embracing virtual goods—such as digital documents and virtual meetings.

In 1994, Peter Menzel's study "Material World", found that families worldwide owned an average of 127 physical objects. If multiplied across 5.6 billion people, that translates to over 700 billion objects. As the world grows wealthier, this number could easily surpass 10s of trillion physical items, increasing emissions through manufacturing, transportation, and waste. However, if these physical items are replaced by digital services—such as virtual art, online education, and telemedicine—we can maintain or even enhance the quality of life while reducing our carbon footprint.

The digital economy's exponential growth offers the possibility of decoupling prosperity from resource consumption. AI and 5G networks will play a pivotal role in this transformation, optimizing systems and enabling smart infrastructure that consumes less energy. Virtual tools and telepresence can reshape industries, from business to education, allowing people to collaborate and thrive without generating unnecessary emissions. Through digital abundance, we can unlock a future where economic growth no longer relies on material expansion but on scalable, sustainable, and healthful digital solutions that benefit everyone.

THE STATE OF ENERGY AND POWER TODAY AND AN UNEVEN WORLD

In the article "The Energy Transition Delusion" (Manhattan Institute) Mark Mills wrote, *"Providing billions of people with cheap and reliable energy has been one of the greatest achievements in history."*

As highlighted in the chart in the article, the share of GDP spent on energy across different periods from the 1300s to the 2000s highlights energy sources' evolution and economic impact over time. It shows how societies shifted their energy dependencies as technologies evolved and new resources were exploited.

1. **Pre-Industrial Era (1300–1800s):**

   - During this period, food, fodder, and wood were primary energy sources. These were essential for human labour, animal labour, and heating or cooking.

   - A large share of the economy was dedicated to energy in the form of food and animal fodder, as most activities relied heavily on physical (human and animal) for agriculture and transportation.

   - Wood was used extensively for heating and as a precursor to early industrial processes. This gradually decreased with the introduction of new fuels.

2. **Industrial Revolution (1800s–1900s):**

- The introduction and rise of coal marked a significant turning point. Coal began to displace wood and significantly reduced the economic burden of energy production by increasing efficiency.

- Oil emerged in the late 19th and early 20th centuries, further enhancing industrial processes and enabling widespread transportation, including automobiles.

3. **20th Century and Modern Era:**

- Gas and non-fossil electricity (hydropower, nuclear, renewables) began to supplement and, in some cases, replace coal and oil.

- Energy as a share of GDP decreased dramatically due to technological innovations, improved energy efficiency, and economic diversification. This shift reflects how industrial economies became less energy-intensive in proportion to their overall financial output.

Primary Energy Use, gigajoules per capita, 2023

- Eastern Africa: 5
- Africa: 14
- Europe: 118
- US: 283

There is a stark disparity in energy consumption between regions. The United States leads with 283 gigajoules per capita of primary energy use in 2023, significantly more than Europe's 118 gigajoules. Africa, as a whole, consumes only 14 gigajoules per capita, while Eastern Africa uses even less at just 5 gigajoules. The data emphasizes that the average American uses 20 times the energy of the average African and 56 times more than someone from Eastern Africa, underlining the slow progress or lack of an energy transition in African regions.

In a year, the sun produces 10^35 Joules total and 3,850,000 EJ to the Earth or 179,000 terawatts (TW) of power every second. This is 10,000 times more energy than humans currently consume.

If global living standards were to rise to current U.S. levels by 2100, energy demand could reach 70-123 TW, depending on population growth.

How Much Energy Will the World Need?

As highlighted by the International Energy Outlook Consumption – By 2050, global energy use in the reference case will increase by nearly 50% compared with 2020—primarily a result of non-OECD economic growth and population, particularly in Asia.

FIGURE 10

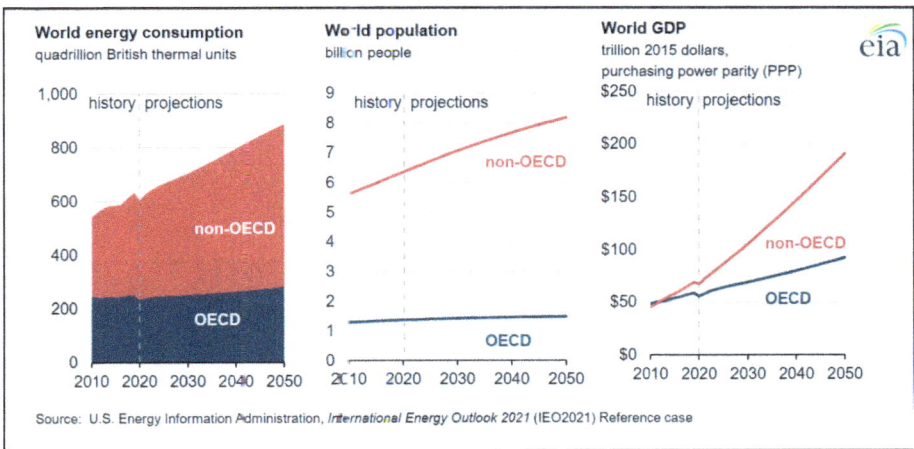

Source: U.S. Energy Information Administration, *International Energy Outlook 2021* (IEO2021) Reference case

AI-powered energy systems can help enhance efficiency, reduce waste, and promote clean, affordable electricity to help achieve the SDG7 Access to Electricity goal[1]. For 92.7% of humanity (7.9 billion people) to have access to affordable, reliable,

---

1   IEA, (2024), 'Access to electricity', accessible at: https://www.iea.org/reports/sdg7-data-and-projections/access-to-electricity (last accessed 26th October 2024)

sustainable, and modern energy by 2030. However, with global access to electricity stalled at 90.2% for the past three years, achieving this target demands shifts in priorities—from conflict and scarcity to using AI-optimized energy systems (optimizing grids, predicting demand, and minimizing waste, energy innovation, and contributing to affordable, reliable, and sustainable electricity).

There is some good news: U.S. energy-related CO_2 emissions decreased by 3% in 2023. The US CO_2 emissions is ~5GT.

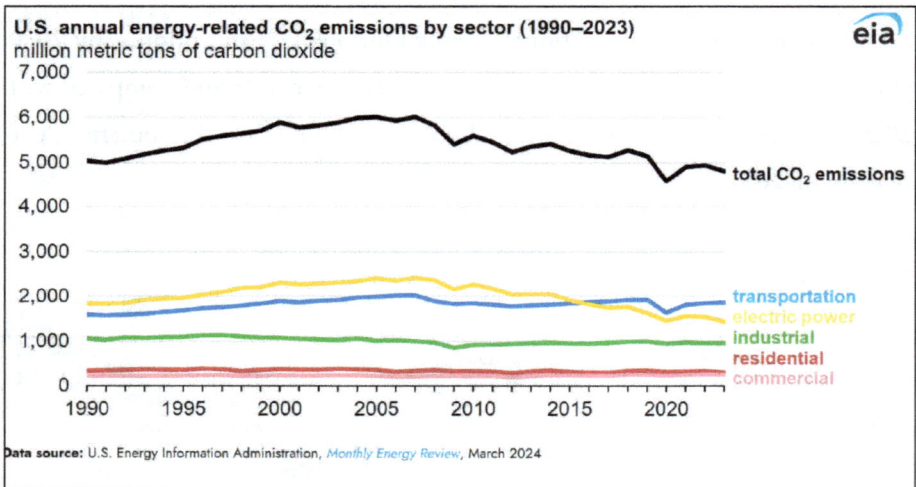

U.S. annual energy-related CO_2 emissions by sector (1990–2023)
million metric tons of carbon dioxide

Data source: U.S. Energy Information Administration, *Monthly Energy Review*, March 2024

GLOBAL ENERGY OUTLOOK: THE PATH TO 2050

The McKinsey Report "The Global Energy Perspective 2023" highlights that global electricity demand could double by 2050, driven by economic growth in emerging markets and widespread electrification. Passenger electric vehicles are expected to achieve cost parity with internal combustion engines by 2025 in major markets like China, Europe, and the U.S., with the global electric car fleet reaching 1.3 billion by 2050.

Solar and wind energy could account for 70% of global installed capacity by mid-century, while gas and other firm resources provide the remaining 30% for reliability. Solar capacity additions are projected to increase significantly, especially in Africa, while wind capacity could grow to 230 GW annually by 2050, requiring a substantial scale-up compared to current levels.

Transmission and distribution (T&D) costs per megawatt-hour could rise significantly, potentially reaching 60–70% of total costs by 2050 in the U.S. and following similar trends in Australia. In scenarios of accelerated decarbonization, increased T&D investments may be partially offset by higher loads from a more electrified economy. Renewable energy costs, such as solar, are projected to drop rapidly, with solar potentially reaching $15–$20 per MWh by 2050.

The 2020s and 2030s are a race to transform energy and provide energy abundance. We shall see if humanity wins this race.

In his article "Seven Reasons To Be Skeptical About SMRs (With Four Charts)" Robert Bryce highlights,

> "Small modular reactors are promising. But commercializing them will be difficult and require staggering amounts of money. Only a few SMR companies will survive. Gas-fired power plants, not nuclear, will meet near-term AI demand.
>
> One of the factors driving the excitement about SMRs is the possibility that they will be used to power Big Tech's data centers and the surge in electricity needed for AI. But the AI boom may be over by the time a substantial number of SMRs get licensed, built, and connected to the grid, which will likely take a decade."

This is not necessarily bad, as this sustainable energy can be used to power the energy needs of the 2030s.

Small Modular Reactors (SMRs) and 3rd/4th Generation Nuclear Reactors: Lowering the Cost of Electricity and CO_2 Emissions for Humanity

The global energy transition hinges on reliable, affordable, and low-carbon electricity technologies. Small modular reactors (SMRs) and advanced third- and fourth-generation nuclear reactors hold transformative potential, offering solutions that simultaneously address economic, environmental, and energy security challenges. By 2050, these atomic advancements could contribute to over 12% of global electricity needs while avoiding billions of tons of carbon dioxide emissions.

Lowering the Cost of Electricity

One of the key barriers to traditional nuclear energy has been its high upfront capital costs. SMRs revolutionize this dynamic with modular construction and factory-based manufacturing, which streamline production, reduce on-site construction times, and lower overall costs. The levelized cost of electricity (LCOE) for SMRs is projected to range between $60 and $100 per megawatt-hour (MWh), making them competitive with natural gas and renewables. For instance, NuScale Power, an industry leader in SMR development, estimates the LCOE for its first-of-a-kind reactor to be approximately $89/MWh, with costs expected to decrease as production scales.

Advanced third- and fourth-generation reactors, such as molten salt and fast neutron reactors, enhance economic efficiency by improving fuel utilization and thermal efficiency. Fourth-generation reactors, capable of achieving up to 45% thermal efficiency compared to the 33% current reactors, can extract more energy per unit of nuclear fuel. Additionally, some designs are projected to reduce the LCOE to below $60/MWh, offering a pathway to cost parity with solar and wind energy.

Lowering CO_2 Emissions

The climate benefits of nuclear energy are substantial. Nuclear power plants prevent approximately 2.5 gigatons of CO_2 emissions annually by displacing fossil fuel-based generation. The introduction of SMRs and advanced reactors could further scale this impact. A study by the International Atomic Energy Agency (IAEA) estimates that deploying SMRs and fourth-generation reactors could result in cumulative CO_2 emission reductions of up to 30 gigatons by 2050, depending on the pace of adoption.

Beyond electricity generation, fourth-generation reactors have the versatility to decarbonize hard-to-abate sectors. High-temperature reactors, for example, can produce industrial heat and hydrogen without CO_2 emissions. The hydrogen generated can then be used as a clean fuel alternative for transportation and heavy industry, sectors currently responsible for nearly 40% of global emissions.

Expanding Access to Clean Energy

SMRs are uniquely suited to bring clean energy to regions where traditional large reactors are impractical. Their compact size and flexible deployment allow them to power remote communities, islands, and industrial operations. By 2035, it is estimated that SMRs could provide electricity to over one billion people in areas without access to reliable grids.

Furthermore, fourth-generation reactors' ability to "burn" nuclear waste reduces the need for long-term storage and addresses one of the public's primary concerns about nuclear energy. This innovation reduces environmental risks and unlocks an additional energy source from spent fuel currently in storage.

Complementing Renewables for Grid Stability

The intermittency of renewables like wind and solar often necessitates backup generation, historically provided by fossil fuels. SMRs and advanced reactors are ideal partners for renewables, offering stable, dispatchable power to maintain grid reliability. In scenarios where renewables dominate the energy mix, nuclear energy can provide the baseload and load-following capabilities necessary to prevent blackouts and stabilize electricity prices.

Global Impact by 2050

By 2050, the IAEA projects that nuclear capacity could expand to 950 gigawatts electric (GWe) under optimistic scenarios, up from the current 390 GWe. This growth would represent over 12% of global electricity generation, significantly contributing to the International Energy Agency's Net Zero by 2050 pathway. SMRs alone could account for 25% of new nuclear capacity, with over 300 reactors potentially deployed worldwide, especially in countries adopting ambitious decarbonization goals.

The cumulative effect of these developments includes:

- Avoiding up to 30 gigatons of CO_2 emissions.
- Generating reliable electricity for billions of people.
- Lowering electricity costs globally, especially in developing nations.

Small modular reactors and third- and fourth-generation nuclear reactors represent a cornerstone of humanity's efforts to achieve a sustainable energy future. Their ability to deliver cost-effective, low-carbon energy aligns with the dual goals of mitigating climate change and ensuring universal access to electricity. As these technologies scale over the next 25 years, they will not only lower the cost of electricity but also drive the global energy transition toward a more abundant and resilient future.

As we look toward 2030, the energy landscape is set to undergo transformative changes driven by innovations in AI, renewables, electric vehicles, and advanced nuclear. These technologies offer hope for a future where energy abundance becomes a reality, unlocking new possibilities for human progress while reducing our environmental footprint. However, achieving this vision will require intentional collaboration across industries, governments, and individuals, as the path to sustainable energy is challenging and guaranteed.

AI will be pivotal in optimizing energy systems, predicting consumption patterns, managing grid reliability, and decoupling economic growth from resource consumption. The increasing reliance on digital infrastructure and virtual solutions will further shift the global economy, enabling prosperity with fewer emissions. Yet, as the world transitions, challenges such as energy poverty, mining for renewables, and uneven access to technology will need to be addressed to ensure equitable progress.

Our decisions today—investing in energy-efficient technologies, supporting policy changes, and advocating for sustainable solutions—will determine whether humanity achieves abundance or faces scarcity.

The future of energy is not just about reducing carbon emissions but also about creating a world where clean, reliable, and affordable energy is accessible to everyone. By embracing AI-driven solutions, expanding renewable infrastructure, and prioritizing innovative policies, we can build a resilient energy system that supports sustainable growth, mitigates climate risks, and fosters a higher quality of life. The race to 2030 has begun. Success will depend on our ability to innovate, collaborate, and commit to a future where energy powers economies and the well-being of humanity and the planet. The choice is ours to make—let's create abundance together.

Education

Abi and Holly were sisters growing up in 2030, navigating a world where school looked nothing like it did even a decade earlier. Abi, a high school senior, and Holly, an eighth grader, were immersed in a learning environment powered by AI, AGI, and immersive tech that made education thrilling and personal. Their days were filled with projects that felt more like real-world adventures than classroom assignments.

Abi was working on her final-year project—designing a sustainable urban garden that could thrive with minimal water. She wasn't doing it alone. Her AI mentor, a digital companion assigned to each student, greeted her daily with insights tailored to her project. Today, the AI mentor suggested a few tweaks based on real-time climate data and even connected her to experts in urban agriculture. Abi's classroom was just as futuristic. Using virtual reality, she and her classmates could walk through a simulation of their garden, testing different layouts, soil types, and irrigation systems before committing to the real thing. For Abi, this wasn't just school—it was practice for her dream of becoming an environmental architect.

Down the hall, Holly was deeply involved in her creative project. Her team was building an interactive storybook that changed its plot and characters based on the reader's emotions. But what made it special was that her team wasn't just from her school; they were students from India, Japan, and Brazil, each bringing their own cultural twists to the story. An AI matching system had paired Holly with like-minded students worldwide, allowing them to work together on the storybook efficiently. Whenever they hit a roadblock, her AI mentor suggested languages, local myths, and even emotional cues they could add to deepen the storyline.

At lunch, the sisters took a few minutes to fill out their digital gratitude journals, a new habit encouraged by their school to keep everyone grounded and emotionally balanced. Mindfulness apps, tailored health tips, and even reminders to stretch or take breaks were part of their routine, helping them manage stress and find joy in the fast-paced, tech-driven environment they were growing up. School was no longer just about academics but about preparing to thrive.

After classes, Abi and Holly walked home, sharing stories from their day. Abi's eyes sparkled as she described the latest VR simulation of her urban garden. Holly laughed, recounting how her team from around the world had gotten into a debate over how to depict a dragon from different cultural perspectives. They were learning more than facts; they were learning how to collaborate, innovate, and create.

As they reached home, Abi turned to Holly with a grin. "Can you believe we're doing this stuff? It's like we're in college already."

Holly grinned back, feeling the same spark. "I know! And I can't wait to see what we'll get to do next year. Imagine the projects we'll be working on by then!"

They laughed, realizing how much they loved learning in this new world. In a future that seemed uncertain for many, they felt confident and excited. With their AI mentors, virtual teams, and the freedom to explore and create, Abi and Holly knew they were building something more than just projects—they were building skills, friendships, and dreams, ready for whatever the future holds.

EDUCATION IN 2030

 # Education 2030

- **Key Themes**
  - Personalized AI mentors
    - Tailored guidance and real-time feedback
    - Adaptive learning to individual needs
  - Project-based learning
    - Real-world problem-solving
    - Multi-disciplinary, collaborative projects
  - Emotional well-being integration
    - Gratitude journals and mindfulness practices
    - Stress management and mental health support
- **Transformative Technologies**
  - Virtual Reality (VR) and Augmented Reality (AR)
    - Immersive classrooms
    - Hands-on VR skills training
  - Blockchain for progress tracking
    - Immutable academic records
    - NFTs as digital certificates
  - AI-powered assessment
    - Continuous evaluation
    - Holistic view of student capabilities
- **Focus on Soft Skills**
  - Emotional intelligence
  - Creativity and critical thinking
  - Teamwork and collaboration
- **Future Learning Modes**
  - Lifelong learning
    - Flexible, modular courses
    - AI-personalized career pathways
  - Gamification
    - Quest-based learning modules
    - Rewards and interactive storytelling
- **Global Connectivity**
  - Virtual classrooms
    - Collaboration across countries
    - Cultural exchange in projects
  - Learning clusters
    - Interest-based groups
    - Global network of learners
- **Why It Matters**
  - Preparing for AI-driven job market
  - Nurturing adaptability and creativity
  - Promoting emotional resilience
- **Impact Areas**
  - Students
    - Tailored learning paths
    - Hands-on, practical skills
  - Teachers
    - Focus on mentoring
    - Facilitating critical thinking
  - Universities
    - Flexible, global learning spaces
    - Skills-based curriculums
  - Governments
    - Digital credentialing policies
    - Lifelong learning initiatives
- **Call to Action**
  - Integrate AI and immersive tech in curriculums
  - Focus on creativity and emotional intelligence

Students will benefit from:

Personalized AI Mentors: By 2030, each student will likely have access to an AI-powered personal mentor. These mentors will use advanced algorithms to analyze a student's learning style, pace, strengths, and weaknesses. They'll provide tailored guidance, adjusting the curriculum in real time to optimize learning outcomes. For example, if a student struggles with a particular math concept, the AI mentor might present the information in a different format or break it down into smaller, more manageable steps.

AI-Powered Assessment: Traditional exams may be replaced by continuous, AI-powered assessments. This system would evaluate students based on their day-to-day performance, project outcomes, and demonstration of skills, providing a more holistic view of a student's capabilities than a single high-stakes exam.

Virtual Classrooms: The concept of a physical classroom will evolve significantly. Virtual Reality (VR) and Augmented Reality (AR) technologies will create immersive learning environments accessible anywhere. Students might attend a history class where they're transported to ancient Rome or a biology class to explore the human body from the inside. This technology will break down geographical barriers, allowing students to collaborate with peers from different countries and cultures.

Project-Based Learning: Education will shift from standardized testing to hands-on, real-world projects that encourage active engagement and practical application. Students will work on multi-disciplinary projects, like designing a sustainable city, which requires skills in environmental science, urban planning, economics, and ethics. Such projects foster critical thinking, teamwork, and problem-solving as students contribute their unique talents while learning to manage tasks collaboratively.

Advanced AI-driven simulations and extended reality will allow students to test their designs virtually, providing real-time feedback on energy use and traffic flow outcomes. This approach will develop students' ability to think holistically, connecting knowledge across subjects, and equip them with adaptable, innovative mindsets essential for today's complex world. Through these immersive projects, education will prepare students to learn and actively apply their knowledge to impact the world.

Focus on Soft Skills: As AI and automation increasingly take on routine tasks, education will profoundly shift to develop uniquely human skills that machines cannot easily replicate. This new approach will prioritize creativity, emotional intelligence, adaptability, and complex problem-solving, aiming to nurture students who can think critically, connect deeply, and flexibly respond to changing environments. Schools will go beyond the basics, offering courses in mindfulness, empathy, resilience, and other emotional and social skills alongside traditional subjects, building well-rounded individuals equipped for the future.

In this evolving educational landscape, a philosophy grounded in the principles of lifelong learning, personal growth, and positive impact promotes a vision of education that is as much about character and purpose as it is about knowledge and skills that prepare students for meaningful lives and fulfilling careers:

- **Perseverance:** Students will learn the value of resilience and grit, developing a mindset that enables them to tackle challenges, learn from failures, and push through obstacles.

- **Creative & Critical Problem Solving:** Creativity and analytical thinking will be integral to the curriculum, allowing students to develop innovative solutions to complex problems.

- **Teamwork and Collaboration:** In a world that values collaboration, students will engage in group projects and team-based challenges, learning to communicate effectively, navigate conflicts, and leverage diverse perspectives.

- **Economic Responsibility:** Courses on financial literacy, resource management, and sustainable economics will teach students how to make wise decisions regarding personal finance and resource allocation.

- **Social Impact and Community Involvement:** Students will be encouraged to see themselves as global citizens with the power to make a difference. Through service-learning projects, internships, and volunteer opportunities, they will apply their skills to real-world social challenges, cultivating a sense of responsibility toward their communities and the world.

- **Environmental Sustainability:** Recognizing the importance of environmental stewardship, students will learn about sustainable practices, climate resilience, and eco-friendly technologies. Hands-on projects in environmental science, renewable energy, and sustainable design will empower students to contribute to a greener, healthier planet.

Education frameworks will be designed to unleash students' potential by focusing on personal growth, purpose, and impact. The curriculum will go beyond academic achievement to help students develop an entrepreneurial mindset—a readiness to identify needs, create solutions, and take action to benefit themselves and others. Mindfulness, empathy, and emotional intelligence courses will help students cultivate self-awareness, manage stress, and build meaningful relationships. Once considered secondary to academics, these qualities will become central to a well-rounded education, preparing students for personal and professional success.

A holistic approach centred on personalized AI learning will inspire students to pursue lifelong learning and equip them to thrive in a world of rapid technological and societal change. This shift will produce skilled workers and adaptable, resilient individuals ready to lead, create, and contribute to a thriving global community.

VR Skills Training: By 2030, VR will revolutionize skills training, providing students with immersive, hands-on experiences that allow them to practice and master complex procedures in a safe, controlled virtual environment. This will be particularly valuable for students pursuing vocational and hands-on careers, enabling them to gain real-world experience without the risks and costs associated with traditional training methods. Students can experiment, make mistakes, and refine their techniques virtually, gaining confidence before entering real-world scenarios.

Here are five key examples of how VR will transform training across various trades and vocational fields: trades (electricians, plumbers, carpenters), pilots, heavy machinery operators, driving (commercial and passenger vehicles), hands-on work (medical procedures, mechanical repair).

Blockchain-Based Progress Tracking: This will provide students with immutable, verifiable records of their academic and skills-based achievements. Unlike traditional systems focusing solely on grades or degrees, blockchain technology will create a comprehensive digital portfolio of a student's educational journey, covering completed courses, projects, skills, certifications, and peer reviews.

NFTs (Non-Fungible Tokens) will serve as digital certificates, marking milestones like project completions, skill mastery, or significant contributions, giving students a lifelong collection of achievements that is easy to share and verify. Additionally, crypto-based incentives could reward students for key accomplishments, encouraging continuous learning. These tokens could be used within educational systems for resources or course access, motivating students to push further.

This blockchain record will be globally accessible and trustworthy, precious for international students or professionals on the move. Based on blockchain-verified records, employers and institutions can assess each student's skills and contributions, such as teamwork or project roles. In this future, blockchain-based records will go beyond traditional transcripts, giving a reliable and transparent view of a learner's real-world skills and achievements, fundamentally transforming how education is valued, and credentials are recognized globally.

Emotional Well-being Integration: Recognizing the importance of mental health, schools will likely integrate emotional well-being into the core curriculum.

This could involve journaling, yoga, deep breathing, regular mindfulness sessions, stress management techniques, and personalized mental health support powered by AI counsellors. Gratitude journals will become essential to school routines, fostering emotional well-being alongside academic development. Daily reflections on positive experiences will help students cultivate mindfulness, reduce stress, develop a stronger sense of empathy, encourage a more optimistic outlook, and improve life, grades and performance.

Lifelong Learning Models: These will reshape education, moving beyond degrees to make learning a continuous journey. Schools and universities will offer flexible, modular courses that allow people to upskill or reskill at any stage of their career, with online, hybrid, and VR options that fit personal schedules and can be accessed from anywhere.

Employers will collaborate with educational institutions to create tailored programs that ensure employees keep up with skills in high-demand areas like AI and cybersecurity, benefiting both the workforce and businesses. AI-powered platforms will personalize learning paths, recommending courses based on individual progress and career goals, making education an ongoing, adaptable process that evolves with each person's life.

Gamification of Learning: Educational content will increasingly be presented in game-like formats to boost engagement. This could involve quest-based learning modules, achievement systems, and competitive elements designed to make learning more enjoyable and motivating.

Students will increasingly form learning clusters based on shared interests rather than geographic proximity. AI-driven platforms will match students with similar passions, allowing them to collaborate on projects, share resources, and learn together, no matter where they live. These small interest-based clusters will create more meaningful and personalized learning experiences, fostering deeper engagement and collaboration. The traditional classroom model will evolve into a network of interconnected learners who can explore topics at a much deeper level, leveraging the global reach of virtual education platforms.

WHAT IS IT?

Education in 2030 will be a transformative blend of technology, personalized learning, and emotional well-being. It will involve personalized AI mentors providing tailored guidance to each student, virtual classrooms accessible from anywhere, and project-based learning that encourages students to solve real-world problems. The focus will be on creativity, critical thinking, and adaptability, supported by VR skills training and blockchain-based progress tracking tools.

WHAT IS GOING TO HAPPEN?

By 2030, education systems will transform into immersive, personalized experiences that center around each student's unique interests and learning style, fostering a global, interconnected approach to knowledge. World-class experts from various fields will participate directly in the education process, offering lectures, workshops, and mentorship through virtual platforms that allow students to learn directly from top minds, regardless of location. This unprecedented access will inspire students to explore specialized fields deeply and build skills grounded in current industry practices and insights from global leaders.

Learning will feel more like a video game than a traditional classroom, with gamified educational systems that motivate students through challenges, rewards, and interactive storytelling. Students will engage in quests to solve math problems, explore history through virtual simulations, or conquer science labs in lifelike VR settings. These gamified platforms will make learning enjoyable, encourage perseverance, and allow students to build essential skills in a format that feels engaging and relevant. Through this gamified approach, students will develop knowledge, resilience, strategic thinking, and teamwork, skills essential for thriving in a future shaped by rapid technological advances.

The shift toward emotional and mental well-being will be integral to the educational experience, with daily practices like gratitude journals and guided mindfulness exercises integrated into the school day. These practices will help students build emotional resilience, manage stress, and cultivate a growth mindset, preparing them to adapt to and thrive in the challenges of modern life. Schools will value emotional intelligence as highly as academic performance, fostering a holistic approach that supports students' mental health and social skills.

Entrepreneurial programs will allow students to create real-world solutions, launch small ventures, or innovate new products, turning their education into a platform for impact. Art, music, and sports will be central as these creative outlets become recognized as vital to developing adaptable, innovative, and well-rounded individuals.

Personalized one-to-one learning through AI mentors will be a cornerstone of education. These AI-powered tutors will adapt to each student's learning pace, style, and preferences, offering real-time feedback, personalized learning paths, and emotional support. The AI mentor will monitor progress, detect areas where the student struggles, and provide targeted resources to bridge knowledge gaps. This approach will replace the traditional one-size-fits-all model, ensuring that every student receives the guidance they need to succeed, regardless of their background or learning challenges. As AI continues to improve, it will become more empathetic and capable of fostering meaningful, human-like student interactions, significantly enhancing engagement and academic outcomes.

Traditional geographical boundaries will dissolve in this new educational landscape. Instead of being limited by the local school district, students will form small, global learning clusters based on shared interests and goals. A student passionate about marine biology in Tokyo might collaborate with peers from California, South Africa, and Brazil, sharing ideas and conducting research as a team. Meeting as avatars in virtual classrooms, accessible anywhere, will create interconnected learning networks, blending cultures, languages, and perspectives to foster a truly global community of learners.

This approach will empower students to pursue purposeful learning, tapping into their interests and creativity and preparing them to contribute meaningfully to an ever-evolving world. By 2030, education will be a journey of discovery, innovation, and connection, nurturing knowledgeable, adaptable, empathetic, and inspired students to make a difference.

WHY DOES IT MATTER?

This evolution is crucial for preparing future generations for a rapidly changing job market, where AI and automation will handle many tasks that humans currently perform. Education must focus on fostering creativity, emotional intelligence, and innovative thinking—skills that machines cannot replicate. The emphasis on mindfulness and well-being will also help create emotionally resilient individuals, reducing stress and promoting lifelong learning habits.

WHO WILL IT IMPACT?

Students will benefit from personalized learning paths tailored to their interests and strengths, allowing them to gain practical skills through hands-on projects. Teachers will shift from traditional lecturing to mentoring, focusing on one-on-one support and fostering critical thinking. Universities will need to adapt by offering flexible, skills-based courses accessible online, transforming campuses into hybrid learning spaces open to global participation.

Industries will gain a workforce with direct experience in creativity, collaboration, and problem-solving, reducing training time and boosting productivity. Meanwhile, governments must adopt policies to support digital credentials, lifelong learning, and equal access, setting standards for a modern education system that values adaptability and continuous growth.

As they respond to these educational changes, governments will face new challenges and opportunities. Policies must adapt, addressing digital credentialing, workforce retraining programs, and lifelong learning initiatives as traditional degrees lose importance.

CALL TO ACTION

Educational institutions must integrate AI, virtual reality, and project-based learning into their curriculums. By doing so, they will prepare students for the technological future and the emotional and creative demands of a world driven by innovation and entrepreneurship.

Finance and Economics

It's a bright, crisp morning in 2030, and Stanley is going through the bustling streets of New York City. As a Harvard graduate working on Wall Street, Stanley has a front-row seat to the digital finance revolution. Meanwhile, his girlfriend, Sunita, who immigrated from India to pursue a PhD in AI, is already deep in her AI research lab in Brooklyn, working on advanced quantum encryption algorithms to secure the next wave of blockchain transactions.

Stanley's morning begins like many others on Wall Street, but it's far from the typical scene you might have imagined a decade ago. The office is buzzing not with frantic brokers shouting over phones but with calm, methodical AI algorithms trading digital assets, from tokenized real estate to fractional shares of renewable energy projects in Brazil. He pulls out his smartphone to check his DeFi portfolio, where AI-powered algorithms are autonomously optimizing his investments across global decentralized finance platforms. Through blockchain-powered distributed finance, Stanley can invest seamlessly in projects worldwide, whether sustainable tech in Kenya or smart cities in Japan. There's no need for banks, brokers, or intermediaries—just direct, frictionless transactions.

On the other hand, Sunita is solving problems at the cutting edge of AI and quantum encryption in her lab. While she sips her chai, she reflects on how much has changed since she moved from India. Gone are the days when transferring money internationally was a headache. Now, thanks to quantum-secure block-chain technology, sending funds back home is as easy as making a phone call, and it's instantaneous. The same encryption technology secures her and Stanley's digital assets—whether their NFTs, cryptocurrencies, or AI-driven savings plan investments.

During lunch, Sunita catches up with Stanley over a video call powered by quantum-enhanced networks. They talk about how different their financial lives are from their parents. For Stanley's father, investing in global markets meant layers of intermediaries, high fees, and waiting days for settling transactions. Now, thanks to quantum-secured distributed finance, they invest globally with just a click and complete trust that their assets are protected by the most advanced encryption on the planet. Sunita remembers how her parents once struggled to get a bank account in India. With blockchain-based digital identities, people anywhere can access financial services without depending on traditional banks.

Later in the afternoon, Stanley receives an alert on his phone: his AI-based financial assistant has rebalanced his portfolio to include more environmental-ly-friendly companies based on his latest preferences. It's an effortless, real-time process without any brokers or fund managers. He's also able to see how his investments in tokenized green bonds are helping finance a new solar farm in South Africa. Meanwhile, Sunita wraps up her day in the lab, knowing that her work in quantum encryption is vital to keeping the world's distributed finance systems secure.

As evening falls, the couple meets at their favorite restaurant, which accepts cryptocurrency payments. They toast a future where distributed finance has made the world more accessible, fair, and secure. With blockchain ensuring transparency, quantum encryption securing their data, and AI optimizing their financial goals, the world of finance has become one without borders, intermediaries, or limits.

For Stanley and Sunita, the future is as seamless as it is exciting—where their investments, savings, and spending are part of a global, secure, and interconnected financial system.

FINANCE and ECONOMICS IN 2030

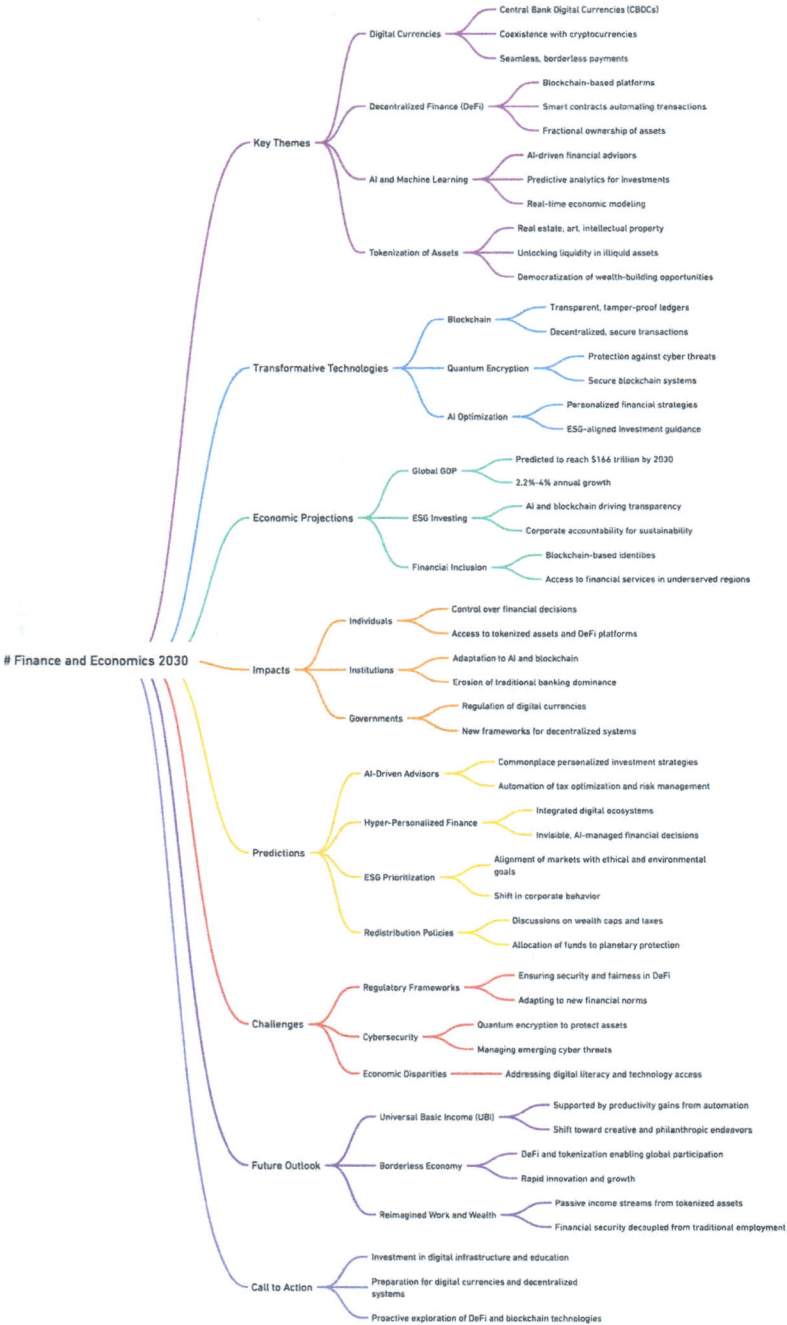

Key Themes
- Digital Currencies
  - Central Bank Digital Currencies (CBDCs)
  - Coexistence with cryptocurrencies
  - Seamless, borderless payments
- Decentralized Finance (DeFi)
  - Blockchain-based platforms
  - Smart contracts automating transactions
  - Fractional ownership of assets
- AI and Machine Learning
  - AI-driven financial advisors
  - Predictive analytics for investments
  - Real-time economic modeling
- Tokenization of Assets
  - Real estate, art, intellectual property
  - Unlocking liquidity in illiquid assets
  - Democratization of wealth-building opportunities

Transformative Technologies
- Blockchain
  - Transparent, tamper-proof ledgers
  - Decentralized, secure transactions
- Quantum Encryption
  - Protection against cyber threats
  - Secure blockchain systems
- AI Optimization
  - Personalized financial strategies
  - ESG-aligned investment guidance

Economic Projections
- Global GDP
  - Predicted to reach $166 trillion by 2030
  - 2.2%-4% annual growth
- ESG Investing
  - AI and blockchain driving transparency
  - Corporate accountability for sustainability
- Financial Inclusion
  - Blockchain-based identities
  - Access to financial services in underserved regions

 # Finance and Economics 2030

Impacts
- Individuals
  - Control over financial decisions
  - Access to tokenized assets and DeFi platforms
- Institutions
  - Adaptation to AI and blockchain
  - Erosion of traditional banking dominance
- Governments
  - Regulation of digital currencies
  - New frameworks for decentralized systems

Predictions
- AI-Driven Advisors
  - Commonplace personalized investment strategies
  - Automation of tax optimization and risk management
- Hyper-Personalized Finance
  - Integrated digital ecosystems
  - Invisible, AI-managed financial decisions
- ESG Prioritization
  - Alignment of markets with ethical and environmental goals
  - Shift in corporate behavior
- Redistribution Policies
  - Discussions on wealth caps and taxes
  - Allocation of funds to planetary protection

Challenges
- Regulatory Frameworks
  - Ensuring security and fairness in DeFi
  - Adapting to new financial norms
- Cybersecurity
  - Quantum encryption to protect assets
  - Managing emerging cyber threats
- Economic Disparities
  - Addressing digital literacy and technology access

Future Outlook
- Universal Basic Income (UBI)
  - Supported by productivity gains from automation
  - Shift toward creative and philanthropic endeavors
- Borderless Economy
  - DeFi and tokenization enabling global participation
  - Rapid innovation and growth
- Reimagined Work and Wealth
  - Passive income streams from tokenized assets
  - Financial security decoupled from traditional employment

Call to Action
- Investment in digital infrastructure and education
- Preparation for digital currencies and decentralized systems
- Proactive exploration of DeFi and blockchain technologies

By 2030, the global economy is anticipated to reach $135-145 trillion annually, based on various growth projections. This represents an approximate growth of 3-4% per year, assuming continued economic expansion, technological advancements, and increasing global integration. The growth rate may fluctuate depending on technological breakthroughs, geopolitical stability, and adopting sustainable industry practices.

The World Bank projects that average global economic growth will slow to 2.2% annually between 2022 and 2031. Using this growth rate and the global GDP figure of $131 trillion in 2019, we can estimate the global GDP in 2030:

Starting with $131 trillion in 2019 and applying 2.2% annual growth for 11 years (2019 to 2030), we get: **$131 trillion * (1.022)^11 ≈ $166 trillion**

Therefore, our best estimate for global GDP in 2030 is approximately $166 trillion. (Perplexity.ai, on Oct 22, 2024)

INTERNATIONAL ENERGY OUTLOOK 2023 – BILLION 2015 DOLLARS TABLE A4. WORLD GDP BY REGION EXPRESSED IN MARKET EXCHANGE RATES

Region	2022	2025	2030	2035	2040	2045	2050
Americas	$28,078	$29,194	$32,220	$35,342	$38,990	$43,033	$47,427
United States	$20,671	$21,361	$23,436	$25,653	$28,414	$31,534	$34,962
Other Americas	$2,516	$2,708	$3,090	$3,496	$3,945	$4,439	$4,976
Europe and Eurasia	$22,949	$23,897	$25,663	$27,295	$29,103	$31,024	$33,091
Asia Pacific	$32,233	$36,392	$44,057	$51,533	$58,587	$65,880	$72,554

Region	2022	2025	2030	2035	2040	2045	2050
Africa and Middle East	$5,526	$6,049	$6,890	$7,795	$8,692	$9,570	$10,432
World	$88,786	$95,533	$108,829	$121,965	$135,372	$149,508	$163,503

INTERNATIONAL ENERGY OUTLOOK 2023 - BILLION 2015 DOLLARS TABLE A3. WORLD GDP PPP

Region	2022	2025	2030	2035	2040	2045	2050
Americas	$32,285	$33,677	$37,275	$40,974	$45,211	$49,889	$54,965
United States	$20,671	$21,361	$23,436	$25,653	$28,414	$31,534	$34,962
Other Americas	$4,273	$4,621	$5,305	$6,050	$6,892	$7,837	$8,882
Europe and Eurasia	$31,730	$33,224	$35,928	$38,541	$41,429	$44,496	$47,822
Asia Pacific	$58,793	$67,172	$83,281	$99,547	$115,494	$132,219	$148,166
Africa and Middle East	$12,838	$14,048	$16,064	$18,211	$20,322	$22,391	$24,401
World	$135,647	$148,121	$172,547	$197,274	$222,456	$248,995	$275,355

WHAT IS IT?

By 2030, global finance will be a hyper-connected, digital-first ecosystem, fundamentally transformed by the convergence of blockchain, artificial intelligence, decentralized finance (DeFi), and digital currencies. The landscape will be defined by blockchain-based decentralized systems, enabling transactions to occur in real time across borders without traditional intermediaries. Smart contracts, tokenized assets, and automated investment platforms will be mainstream, offering individuals unparalleled control and flexibility over their financial lives.

Central Bank Digital Currencies (CBDCs) will be adopted by over 80% of the world's economies, providing governments with more effective tools for monetary policy while coexisting with cryptocurrencies like Bitcoin and Ethereum. Traditional banking systems will be reimagined, seamlessly integrating with fintech innovations and DeFi platforms to provide more inclusive, accessible financial services. Predictive analytics and AI-driven platforms will optimize investments, with algorithms capable of analyzing millions of data points to forecast market trends and manage risks at scales far beyond human capability.

Adopting quantum-resistant encryption will safeguard global transactions, ensuring financial security against emerging cyber threats. Blockchain technology will create transparent, tamper-proof ledgers for real-time monitoring of financial markets, eliminating inefficiencies and bolstering trust in global systems. Tokenized real-world assets, including real estate, commodities, and equities, will unlock trillions of dollars in liquidity, allowing fractional ownership and democratizing access to investments.

In this new era, data-driven decision-making will dominate. Individuals, businesses, and governments will rely on AI-powered financial advisors capable of crafting hyper-personalized strategies that optimize wealth growth, manage risks, and align with ethical and sustainability goals. As these technologies converge, finance will become more efficient and inclusive and a driving force for global economic stability and equitable growth.

WHAT IS GOING TO HAPPEN?

Global finance will experience significant shifts by 2030. The rise of digital currencies, private (e.g., cryptocurrencies) and public (CBDCs), will redefine how money is exchanged and managed. Automated, AI-driven investment strategies will continue to grow, providing individuals and institutions with personalized financial advice in real time. Blockchain technology will underpin most economic transactions, ensuring transparency, security, and accountability. The integration of quantum-resistant encryption will protect financial networks from quantum computer threats. Additionally, new regulations will be required to oversee the convergence of traditional finance, decentralized platforms, and the growing role of AI in the decision-making process.

In 2030, national currencies and traditional banks will coexist with myriad digital assets and decentralized financial systems. Governments will issue Central Bank Digital Currencies (CBDCs) to maintain monetary control, but the rise of stablecoins, cryptocurrencies, and tokenized assets will provide alternative means of value storage and exchange. The barriers to global trade and investment will be lowered, allowing anyone with internet access to participate in the worldwide economy. AI-driven advisors will manage investment portfolios, optimizing for personalized goals and risk tolerance. At the same time, peer-to-peer lending and decentralized insurance platforms will flourish, cutting out the intermediaries and reducing costs.

We will also witness the maturation of ESG (Environmental, Social, and Governance) investing. Fueled by AI and blockchain transparency, financial markets will prioritize sustainability and ethical practices. Companies must align with these values to attract investment, leading to a profound shift in corporate behaviour and global economic priorities.

WHY DOES IT MATTER?

This transformation will democratize access to financial services, allowing individuals worldwide to participate in the global economy with fewer barriers. It will streamline cross-border transactions, reduce the time and cost of moving capital, and improve financial inclusion. The rise of decentralized finance platforms will give people more control over their assets, bypassing traditional financial institutions. However, these changes will also present challenges, such as the need for new regulatory frameworks and the potential for increased cybersecurity risks in an increasingly digitized financial landscape.

The financial landscape of 2030 will democratize access to wealth-building opportunities, allowing billions of people worldwide to participate in the global economy. This shift will erode traditional power structures in finance, where large institutions have historically held sway. With decentralized systems, individuals will have greater financial autonomy, while smart contracts and AI will reduce the need for intermediaries, lowering costs and increasing efficiency. However, this will also bring challenges, including the need for robust cybersecurity, new regulatory frameworks, and the potential for economic disparities in digital literacy and access to technology that still need to be addressed.

WHO WILL IT IMPACT?

Global finance in 2030 will impact various stakeholders, from governments and financial institutions to everyday consumers. Governments must adapt to the shift toward digital currencies, regulating and securing financial systems against new threats like quantum hacking. Banks and financial institutions will be forced to innovate, leveraging AI and blockchain to stay competitive. Consumers will benefit from faster, cheaper, and more accessible financial services, though they will also need to navigate new technologies and the risks they bring.

This transformation will impact everyone—from individuals managing their finances to large corporations navigating a rapidly changing economic landscape. Entrepreneurs will benefit from easier access to capital through tokenized fundraising and decentralized markets. Investors will have a broader range of assets to choose from, with real-time insights powered by AI. Traditional financial institutions must adapt or risk obsolescence, while governments will grapple with the challenge of regulating a borderless, digital economic system. On a macroeconomic scale, emerging markets could experience unprecedented growth as they leapfrog traditional banking infrastructure, while developed economies must innovate to stay competitive.

CALL TO ACTION

Governments, businesses, and individuals must invest in financial education and digital infrastructure to keep pace with the evolving landscape of global finance. Preparing for the rise of digital currencies, AI-driven investments, and decentralized financial systems will ensure that people can fully participate in and benefit from the future economy. Regulatory frameworks must evolve to provide security and trust in this rapidly changing environment.

Start by educating yourself about decentralized finance (DeFi) and digital currencies. Open a digital wallet, explore blockchain-based financial services, and invest in a cryptocurrency or tokenized asset. Understanding these emerging technologies today will give you a significant advantage as they become the foundation of finance in 2030.

PREDICTIONS FOR 2030

Decentralized Finance (DeFi) Will Be Mainstream: By 2030, decentralized finance (DeFi) platforms will rival traditional financial institutions in scale and influence. People will manage their finances on blockchain-based platforms, accessing loans, savings, investments, and insurance without needing a bank. This shift will eliminate barriers like high fees, limited access, and slow transactions, making financial services more inclusive and efficient. Smart contracts will automate transactions, ensuring transparency and reducing fraud. Imagine instantly transferring money to a friend overseas, with no middlemen or fees, or securing a loan within minutes, collateralized by digital assets.

AI AND MACHINE LEARNING IN FINANCE

The financial industry is poised for a significant transformation. The rapid advancements in Artificial Intelligence (AI) and Machine Learning (ML) have created unprecedented opportunities for innovation and disruption. In this chapter, we will explore the role of AI and ML in shaping the global finance landscape in 2030, examining the industry's current state, the challenges and opportunities that lie ahead, and the potential impact on financial institutions, markets, and individuals.

In recent years, AI and ML have made significant inroads in the financial industry, with applications ranging from risk management and portfolio optimization to customer service and fraud detection. Some of the key areas where AI and ML are being used include:

- **Risk Management:** AI-powered systems analyze large datasets and identify potential risks, allowing financial institutions to make more informed decisions.

- **Portfolio Optimization:** ML algorithms are being used to optimize investment portfolios, considering factors such as market trends, economic indicators, and client preferences.

- **Customer Service:** Chatbots and virtual assistants are being used to provide 24/7 customer support, freeing up human customer service representatives to focus on more complex issues.

- **Fraud Detection:** AI-powered systems detect and prevent fraudulent activities like identity theft and money laundering.

Digital Currencies Will Coexist with Traditional Money: National currencies will be digital, issued as Central Bank Digital Currencies (CBDCs), coexisting with cryptocurrencies and stablecoins. Your digital wallet will hold a mix of government-backed currency, crypto, and tokenized assets, all interchangeable.

Payments will be seamless, instant, and borderless. Whether buying coffee at a local café or investing in a global startup, transactions will be as simple as a tap on your device or a voice command. Traditional bank accounts will become relics, with financial services embedded directly into social platforms, e-commerce sites, and virtual worlds.

AI-driven Financial Advisors Will Be Commonplace: The financial advisory industry will be dominated by AI-powered platforms that provide personalized investment strategies, tax optimization, and risk management. These platforms will continuously analyze market trends, personal goals, and even behavioural patterns to make real-time adjustments to your portfolio. For example, your AI advisor might notice a spike in your spending on travel and suggest cutting back on discretionary purchases or reallocating investments to fund your next big trip. Financial planning will be dynamic, responsive, and tailored to individual needs, ensuring everyone, not just the wealthy, can access sophisticated financial advice.

Tokenization of Assets Will Redefine Ownership: By 2030, almost any asset, from real estate to art to intellectual property, will be tokenized and tradable globally. This will unlock liquidity in traditionally illiquid assets, allowing fractional ownership and greater diversification. For instance, you could own a fraction of a Picasso painting or a share of a high-rise in Manhattan through digital tokens. This democratization of wealth-building opportunities will reshape investment strategies, enabling more people to benefit from assets previously accessible only to the wealthy. This also means that your digital portfolio might include traditional stocks and tokenized shares of various real-world assets, providing a more diverse and resilient financial foundation.

Financial Inclusion Will Be Revolutionized: The barriers that have historically kept billions of people out of the financial system will be dismantled. Blockchain technology will provide secure, verifiable identities, allowing even those without traditional identification to access financial services. Microloans, insurance, and savings products will be tailored to underserved populations and delivered via mobile platforms. Imagine a farmer in a remote village using a smartphone to secure crop insurance or a young entrepreneur in a developing nation raising funds through a global decentralized platform. The economic impact of bringing billions into the financial fold will be transformative, driving international growth and reducing inequality.

90% Tax on All Wealth Over $1B: Global discussions may exist around imposing a hefty tax (as high as 90%) on all wealth over $1 billion to reduce extreme wealth concentration. Such measures could lead to significant reductions in wealth inequality, with the additional tax revenue used to fund public services, climate initiatives, and global poverty reduction efforts. This tax policy, while controversial, would be a return to past taxation schemes that may be seen as a way to redistribute wealth more fairly and address global social and economic disparities.

2% of the Global Economy Towards Planetary Protection: In addition to 2% of GDP spent on defence budgets, governments and corporations may agree to allocate 2% of the global economy toward planetary protection initiatives, focusing on climate change mitigation, environmental sustainability, and renewable energy development. This funding would be channelled into large-scale projects to reduce carbon emissions, protect biodiversity, and develop clean technologies, ensuring that the global economy supports long-term planetary health. This collective effort would be crucial to addressing the urgent challenges of climate change and environmental degradation.

PREDICTIONS FOR 2040

Hyper-Personalized Financial Ecosystems: By 2040, your financial life will be completely integrated with your digital presence. AI-driven ecosystems will manage your finances across all aspects of your life, from daily expenses to long-term investments, tailored to your goals, values, and preferences. For example, your system might automatically adjust your investment portfolio based on your carbon footprint or divert funds into a health savings account when you increase your exercise routine. Financial decisions will become nearly invisible as the AI adapts and learns from your lifestyle in real time.

The Global Economy Will Become Truly Borderless: The convergence of DeFi, digital currencies, and tokenization will create a truly global economy where capital, talent, and innovation flow freely across borders. The idea of national economies will blur as value is created, exchanged, and stored in decentralized networks, accessible to anyone with an internet connection. This borderless economy will be powered by AI and data, allowing for rapid innovation and economic growth on an unprecedented scale. Startups will no longer be constrained by geography, and consumers will easily access goods, services, and investment opportunities from around the world.

The Reimagining of Work and Wealth: As automation, AI, and robotics take over more jobs, the concept of work and wealth will evolve. Universal Basic Income (UBI) or similar models will likely be in place, supported by technological productivity gains. People will spend less time working for money and more time pursuing creative, entrepreneurial, or philanthropic endeavours. Wealth will no longer be solely tied to labor as passive income streams from tokenized assets, royalties, and automated investments become more prevalent. This shift will lead to a more balanced, purpose-driven society where financial security is decoupled from traditional employment.

The future of finance and economics in 2030 is not just about new technologies or investment strategies but a fundamental shift in how we interact with money, wealth, and opportunity. As the world becomes more interconnected, decentralized, and digital, the power to shape your financial destiny will be in your hands like never before.

Inflation in 2030

In 2030, inflation will be more dynamically managed through advanced AI-driven economic models. Central banks and governments will have access to real-time data, allowing quicker adjustments to monetary policy in response to inflation-ary pressures. The traditional 2% inflation target will still guide economic policies in many countries. Still, the tools to control it will be more refined, potentially allowing inflation to remain lower in some regions. However, global factors such as climate change, commodity shortages, and geopolitical tensions may lead to inflation spikes in specific sectors, making inflation management a critical and complex task.

In a world where inflation is managed with AI-driven economic models and real-time data, Bitcoin and other cryptocurrencies will emerge as powerful decentralized hedges against inflationary risks. While central banks refine their tools to stabilize inflation around traditional targets, external pressures like climate change, com-modity shortages, and geopolitical tensions will continue to cause localized spikes. In this environment, cryptocurrencies offer an alternative to centralized monetary systems, protecting individuals and businesses from currency devaluation. With their fixed supply (as seen with Bitcoin) and global accessibility, they provide a store of value and financial sovereignty, empowering users to safeguard their wealth amid economic uncertainty. Decentralization will prove to be a vital force in navigating an increasingly volatile financial landscape.

Mobility

"The itch that led our ancestors to risk everything to travel in small boats across large bodies of water like the Pacific Ocean is related to the drive that will one day lead us to colonize Mars. Its origins lie in a mixture of culture and genetics." —**Dr. Chris Impey, *Beyond: Our Future in Space*** *

The hum of electric vertical takeoff and landing vehicles (eVTOLs) filled the sky over a bustling metropolis as Sam, the ambitious CEO of an up-and-coming flying drone taxi company, oversaw another successful morning of operations. His company's fleet of autonomous air taxis had become a lifeline for busy executives, offering a seamless solution to inner-city travel. These sleek machines effortlessly bypassed the gridlocked streets below, promising both convenience and a more sustainable future for urban mobility.

Across town, Anil, a private equity investor and early supporter of autonomous transportation, began his day in the comfort of his automated home. Anil's day revolved around precision—leveraging data and efficiency to make high-stakes investment decisions. His commute involved a combination of autonomous vehicles tailored to his schedule. A self-driving electric car picked him up from his driveway, offering a quiet moment for reviewing his morning

---

* Used with permission

briefings before connecting him to one of Sam's eVTOLs. The ride soared above the congested avenues, delivering him to his meeting in minutes.

Their paths first crossed when Anil recognized the transformative potential of Sam's technology. Years earlier, he had been one of the initial investors in the flying drone taxi company. However, their collaboration went beyond financial support. Together, they envisioned a future where transportation wasn't just a service for the elite but an equitable, sustainable solution for everyone.

For Sam, it was about more than innovation—it was about impact. His fleet of drones wasn't just ferrying executives; they were also being used to deliver essential medical supplies to underserved areas. By the early 2030s, eVTOLs had become integral to disaster response, capable of transporting aid to remote locations more quickly. "Transportation is a human right," Sam often said. "If we can reimagine urban mobility, we can redefine who has access to opportunity."

Anil's role, however, was strategic. As the technology gained traction, he pushed for policies that subsidized these services for middle-class workers, ensuring that the benefits of next-generation transportation reached beyond corporate boardrooms. For him, each meeting wasn't just about profit margins—it was a chance to advocate for a future where accessibility and sustainability went hand in hand.

Their shared vision materialized in a city transformed. Congested streets once choked with emissions now thrived with greenery, thanks to reduced car usage. Reclaimed from parking lots and highways, public spaces flourished with vibrant parks and walking trails. Anil's investments and Sam's operational ingenuity proved that bold ideas could create lasting change.

One evening, over drinks at a rooftop bar illuminated by the glowing lights of their aerial fleet, Anil looked at Sam and said, "You've built something extraordinary here. But what's next?" Sam smiled, gazing at the fleet soaring above the skyline. "Next? We're taking it global. There are millions of people in cities across the world who need this. Mobility isn't just about getting from point A to B. It's about unlocking potential—giving people time, access, and opportunity."

And as the two visionaries toasted, they knew their work had only begun. Their partnership wasn't just reshaping transportation; it paved the way for a world where technology worked for everyone, driving humanity forward with every lift-off.

TRANSPORTATION IN 2030

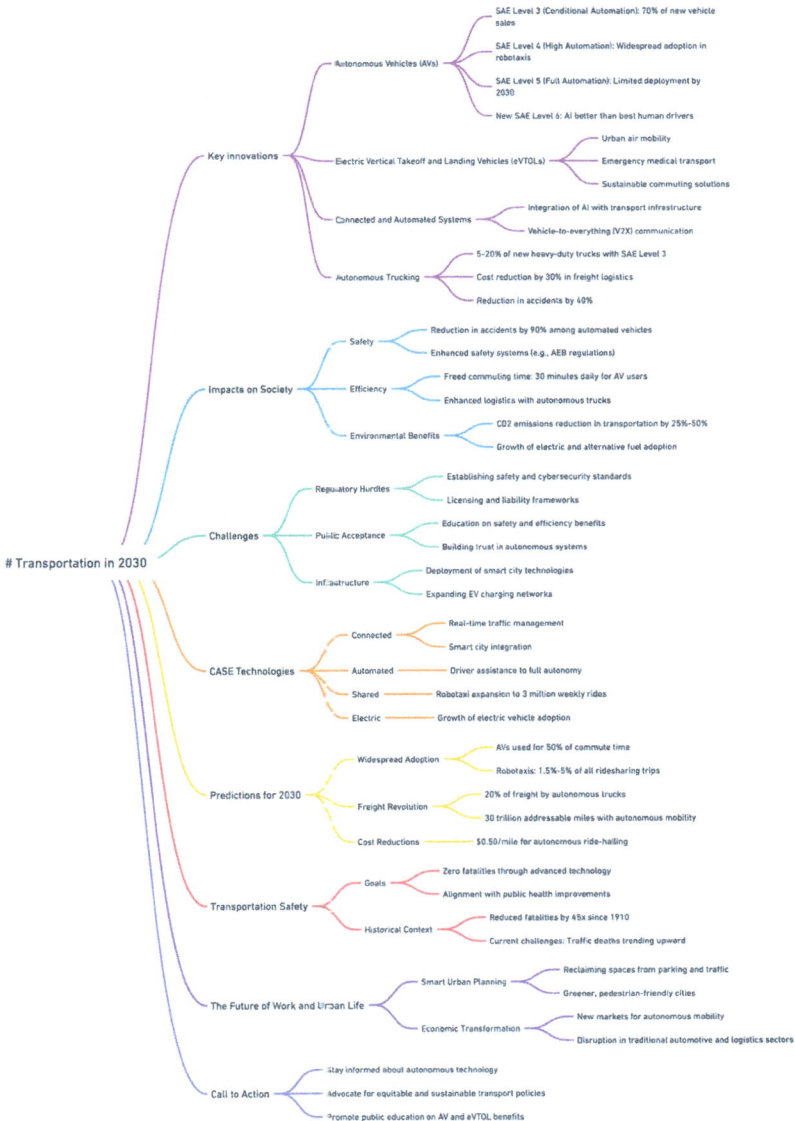

Transportation in 2030

- **Key Innovations**
  - Autonomous Vehicles (AVs)
    - SAE Level 3 (Conditional Automation): 70% of new vehicle sales
    - SAE Level 4 (High Automation): Widespread adoption in robotaxis
    - SAE Level 5 (Full Automation): Limited deployment by 2030
    - New SAE Level 6: AI better than best human drivers
  - Electric Vertical Takeoff and Landing Vehicles (eVTOLs)
    - Urban air mobility
    - Emergency medical transport
    - Sustainable commuting solutions
  - Connected and Automated Systems
    - Integration of AI with transport infrastructure
    - Vehicle-to-everything (V2X) communication
  - Autonomous Trucking
    - 5–20% of new heavy-duty trucks with SAE Level 3
    - Cost reduction by 30% in freight logistics
    - Reduction in accidents by 40%
- **Impacts on Society**
  - Safety
    - Reduction in accidents by 90% among automated vehicles
    - Enhanced safety systems (e.g., AEB regulations)
  - Efficiency
    - Freed commuting time: 30 minutes daily for AV users
    - Enhanced logistics with autonomous trucks
  - Environmental Benefits
    - CO2 emissions reduction in transportation by 25%–50%
    - Growth of electric and alternative fuel adoption
- **Challenges**
  - Regulatory Hurdles
    - Establishing safety and cybersecurity standards
    - Licensing and liability frameworks
  - Public Acceptance
    - Education on safety and efficiency benefits
    - Building trust in autonomous systems
  - Infrastructure
    - Deployment of smart city technologies
    - Expanding EV charging networks
- **CASE Technologies**
  - Connected
    - Real-time traffic management
    - Smart city integration
  - Automated
    - Driver assistance to full autonomy
  - Shared
    - Robotaxi expansion to 3 million weekly rides
  - Electric
    - Growth of electric vehicle adoption
- **Predictions for 2030**
  - Widespread Adoption
    - AVs used for 50% of commute time
    - Robotaxis: 1.5%–5% of all ridesharing trips
  - Freight Revolution
    - 20% of freight by autonomous trucks
    - 30 trillion addressable miles with autonomous mobility
  - Cost Reductions
    - $0.50/mile for autonomous ride-hailing
- **Transportation Safety**
  - Goals
    - Zero fatalities through advanced technology
    - Alignment with public health improvements
  - Historical Context
    - Reduced fatalities by 45x since 1910
    - Current challenges: Traffic deaths trending upward
- **The Future of Work and Urban Life**
  - Smart Urban Planning
    - Reclaiming spaces from parking and traffic
    - Greener, pedestrian-friendly cities
  - Economic Transformation
    - New markets for autonomous mobility
    - Disruption in traditional automotive and logistics sectors
- **Call to Action**
  - Stay informed about autonomous technology
  - Advocate for equitable and sustainable transport policies
  - Promote public education on AV and eVTOL benefits

PREDICTIONS FOR 2030

- **Automated Driving Adoption:** 70% of NEW vehicle sales will support Automated Driving (SAE Level 3 Conditional Automation- Eyes Off but the driver will need to be ready to engage when required), with drivers using the AI for 50% of their commute time, allowing for activities like reading or working, saving an estimated 30 minutes of a person's day to do other things while in the car. Auto manufacturers will provide liability for accidents occurring in Level 3, where the vehicle handles driving in specific conditions, with the driver ready to intervene if requested.

- **Reduced Traffic Accidents and Insurance Costs:** Traffic accidents and fatalities will decrease by 90% among vehicles equipped with Level 3 technology, leading to significantly lower insurance premiums. In contrast, those without this technology will face increased insurance costs.

- **Growth in Robotaxi Rides:** The number of Robotaxi rides will exceed three million weekly (using Level 4 SAE High Automation where the car can operate without human intervention), accounting for over 1.5% and as high as 5% of all ridesharing trips. This growth is driven by the rapid expansion of services like Waymo, which in 2024 has already achieved 150,000 Autonomous Ride-Hailing trips per week.

- **Widespread Adoption of Autonomous Trucks:** 5-20% of new heavy-duty trucks will be equipped with SAE Level 3 automation, enabling fully autonomous highway operation without human intervention. This advancement will significantly enhance long-haul freight efficiency.

- **Reduction in Operational Costs:** The integration of autonomous trucks is projected to decrease total operating costs by over 30%, primarily through savings on labor and fuel. This reduction will make freight transportation more cost-effective and competitive. BCG

- **Enhanced Safety and Reduced Accidents:** The deployment of autonomous trucks could lower accident rates by 40%, as these vehicles can operate without fatigue and are equipped with advanced safety systems, leading to safer highways.

These levels, defined by the Society of Automotive Engineers (SAE), range from Level 0 (no automation) to Level 5 (full automation):

1. **Level 0 (No Automation):** The driver controls everything.

2. **Level 1 (Driver Assistance):** Basic features like cruise control; the driver is still in control.

3. **Level 2 (Partial Automation):** The car can control steering and speed, but the driver must remain engaged.

4. **Level 3 (Conditional Automation):** The car can handle most driving tasks, but the driver must be ready to intervene.

5. **Level 4 (High Automation):** The car can operate without human intervention in certain conditions or areas, but there might be areas where it hands control back.

6. **Level 5 (Full Automation):** No human driver is needed; the car can handle all driving tasks under all conditions.

I am defining a new Level 6 where the AV/AI is better than the best driver, and only an AI can operate in the environment.

PREDICTIONS FOR 2040

- **90% of ALL vehicles** will support Automated Driving (SAE Level 3 or greater), with drivers engaging it for 90% of their commute time, allowing for activities like reading or working, saving an estimated 55 minutes daily to do other things while in the car.

- **Autonomous Vehicles:** 20% of new vehicles sold will be SAE Level 4/5, allowing nearly full automation during commutes and daily travel.

- **50% Reduction of all Accidents:** Automated/Autonomous technology will dramatically lower fatalities and insurance costs.

- **100 Million Weekly Robotaxi Rides:** Robotaxi services will expand, accounting for 50% of all global ridesharing trips.

- **Dominance of Autonomous Freight Transport:** By 2040, autonomous trucks will handle 60% of all freight transportation, with a significant portion operating without human drivers, optimizing logistics and supply chain operations.

- **Further Reduction in Accidents:** The continued advancement and integration of autonomous technology are anticipated to result in a 50% reduction in truck traffic accidents, contributing to overall road safety.

- **Environmental Benefits:** The widespread use of autonomous trucks, many of which will be electric or utilize alternative fuels, is expected to reduce carbon emissions from freight transport by 25%, supporting global sustainability goals.

AGI (Artificial General Intelligence) will arrive for transportation when 50% of vehicles sold can operate at Level 4 SAE High Automation (where the car can operate without human intervention) at a cost of less than $20K, which is less than $0.40 per mile. Traffic accidents and fatalities will decrease by 90% for these vehicles. Overall, road deaths decrease by <50%. CO_2 emissions have dropped by 50% for the transportation sector.

ASI (Artificial Super Intelligence) will arrive for transportation when 90% of vehicles sold can operate at Level 5 SAE Full Automation: (No human driver is needed; the car can handle all driving tasks under all conditions) at a cost of less than $10K, which is less than $0.20 per mile. Traffic accidents and fatalities will decrease by 99% for these vehicles. Overall, road deaths will decrease by <90%. Five years later, there are no road deaths (from the current 1.3M road deaths today). CO_2 emissions will have dropped by 90% for the transportation sector.

In 2024, the total number of taxi and ridesharing trips globally per week is 200 to 300 million.

WHY IT MATTERS?

The rapid evolution of transportation technology, particularly in AI and autonomous vehicles, has the potential to reduce road fatalities and drastically enhance mobility. This shift could transform how we live, work, and interact, making transportation safer, more efficient, and sustainable. It could free up one hour a day of a commuter's time to do something other than driving. However, it also presents challenges, including potential risks and the need for new regulations.

WHO WILL IT IMPACT?

This transformation will impact everyone, from daily commuters to policymakers. Industries like automotive, insurance, and urban planning will undergo significant changes, while individuals will experience personal transportation and lifestyle shifts.

CALL TO ACTION

Stay informed about advancements in autonomous vehicle technology and AI, and advocate for regulations prioritizing safety, affordability, and environmental sustainability. Know how to use the technology responsibly and safely.

DOUG HOHULIN ON TRANSPORTATION

Autonomous and Connected Vehicle leadership roles and strategic engagements at Nokia have shaped my journey in transportation technology. This expertise extends through representations at Department of Transportation (DoT) workshops and collaborations with influential entities such as the 5G Automotive Association (5GAA), CTIA, and the GSM Association (GSMA).

A key aspect of my expertise was my membership in the 5GAA FCC Task Force, analyzing the Infrastructure Investment and Jobs Act (IIJA) to enhance Vehicle-to-Everything (V2X) communications and engaging in high-level advocacy, promoting 5GAA's goals within the federal government.

In urban planning, I collaborated with Infrastructure Owners and Operators (IOO) (road operators) to integrate Automated Vehicles/Connected Vehicles (AV/CV) and Smart City Technologies into municipal frameworks. This involved responding to RFI/Ps from cities and counties, helping them prepare for the technological shifts in transportation. I contributed to serve on cooperative automated transportation working groups. I was involved in state and regional task forces like the Texas CAV Task Force and the Kansas City Metro AV Task Force. These roles allowed me to influence transportation policies and practices to align with cutting-edge technologies. My editorial contributions to the 5GAA regulatory responses and white papers have helped shape industry standards and guidelines for evolving innovative transportation solutions. As we approach 2030, my continued involvement aims to drive innovation in transportation, ensuring smarter, more connected cities and a safer road system and the elimination of over 1.2-1.3 million road deaths each year.

WHAT IS IT?

Transportation has been a fundamental driver of human progress, evolving from simple biological modes like walking to today's complex technological systems. This chapter explores the "8Gs of Transportation," tracing the journey from early human locomotion through maritime, rail, automotive, and aviation revolutions to the emerging frontiers of autonomous vehicles.

As we stand on the cusp of a new transportation era, the potential for transformative change is immense. Automated and autonomous vehicles promise to reshape our cities, improve safety, and redefine mobility. Yet these advancements also bring challenges, from regulatory hurdles to ethical considerations.

The impact of transportation on society cannot be overstated. It influences everything from urban development and economic growth to public health and environmental sustainability. As we look towards 2030 and beyond, integrating artificial intelligence, connectivity, and electrification in transportation systems can dramatically reduce accidents, congestion, and emissions.

This chapter will examine current trends, future predictions, and the broader implications of transportation innovations. By understanding where we've been and where we're headed, we can better navigate the road ahead and shape a transportation future that is safer, more efficient, and more sustainable for all.

A Brief History of Transportation and the 8Gs of Transportation

Transportation has dramatically evolved from the simple act of walking to the sophisticated realm of space travel. This journey can be categorized into eight distinct generations, each marked by groundbreaking innovations that revolutionized how humans traverse the globe and beyond.

- **0G: Biological** The most basic form of transportation is biological, encompassing human walking and the domestication of animals for riding. Early humans relied on their two feet to explore their surroundings, hunt, and gather resources.

- **0.5G: Riding Animals** Expanding upon human walking and riding animals such as horses, camels, and elephants allowed ancient civilizations to trade, wage wars, and explore unknown territories over vast distances. This phase in transportation was pivotal in shaping early human societies and their interactions. Taming animals like horses significantly increased their travel speed and range, marking the first enhancement of human mobility.

- **1G: Ships & Boats** The advent of ships and boats, with archaeological evidence suggesting their use around 10,000 years ago, enabled humans to explore and connect across water bodies. Maritime transportation was instrumental in the spread of cultures, the establishment of trade routes, and the discovery of new lands.

- **2G: Trains** The introduction of the steam train by George Stephenson in 1814 marked a revolutionary shift in land transportation. Trains enabled more efficient movement of goods and people across continents, fueling industrial growth and urbanization. The railway system became the backbone of industrial societies.

- **3G: Automobiles** Automobiles, introduced by Carl Benz in 1886, brought personal mobility to the masses. The development of the automobile industry led to the expansion of roads and highways, reshaping landscapes and enabling suburban lifestyles. Cars have become symbols of freedom and economic status.

- **4G: Planes** The Wright brothers' first flight in 1903 launched the era of air travel. Airplanes transformed the 20th century, making crossing oceans and continents in hours instead of days or weeks possible. This generation of transportation shrunk the world, facilitating global business and international relations.

- **5G: Automated & Autonomous Vehicles** The ongoing development of automated and autonomous vehicles represents a significant leap towards safer and more efficient road travel. These technologies aim to reduce human error, the leading cause of traffic accidents, and optimize traffic flow, making transportation more sustainable.

- **6G: Virtual** Moving from physical to virtual transportation, this era focuses on communication technologies that eliminate the need for physical travel. Video conferencing and virtual reality have begun to replace business trips and meetings, reflecting a shift towards a more digital, interconnected world.

- **7G: Electric Take off and Landing Vehicles (eVTOLs)** eVTOLs represent the next generation of urban and regional air mobility, designed for efficient, autonomous passenger and cargo transport. They feature enhanced connectivity, real-time data processing, and optimized flight safety. Key advancements include AI-based flight control systems, low-noise electric propulsion, and seamless integration with smart city infrastructure. With benefits like reduced emissions and faster travel, eVTOLs aim to revolutionize urban transit by offering quick, eco-friendly alternatives to road congestion. Scalable operations are also expected as regulatory frameworks evolve to accommodate aerial mobility networks.

- **8G: The Final Frontier – Space** The final frontier of transportation extends beyond Earth. Innovations in space travel aim to make extraterrestrial journeys a reality, with implications for space colonization and interplanetary travel. Each generation of transportation has played a pivotal role in shaping human civilization and its progression. As we look toward the future, transportation boundaries continue to expand, promising new levels of connectivity and exploration.

In 2010, One Billion Vehicles on Earth

Reaching the landmark figure of one billion vehicles in 2010 has significantly transformed society, impacting everything from urban development to global environmental health. This monumental increase in cars has revolutionized personal and commercial mobility, enabling greater freedom of movement and facilitating economic interconnectivity on an unprecedented scale. Cities and suburbs have expanded, with infrastructure evolving to accommodate the growing reliance on personal transportation, leading to profound changes in lifestyle and community structures.

Unlocking the Market Potential of Autonomous Mobility: Cutting the Cost of Transporting Everything

As highlighted by the "Autonomous Mobility: Examining the Economics White Paper" by ARK invest (ark-invest.com), the future of transportation is poised for a revolution as advancements in artificial intelligence (AI) and electric battery technology reshape how people and goods move. For over a century, transportation costs have remained stagnant, with the average cost of vehicle ownership and operation fixed at around $0.70 per mile since the days of the Model T. However, autonomous electric vehicles (EVs) promise to drastically reduce costs, making mobility more accessible, efficient, and scalable. ARK Invest projects that these innovations will transform both passenger transportation and freight logistics, creating economic opportunities at an unprecedented scale.

At the heart of this transformation lies the ride-hailing market, which could grow from $300 billion in value today to over $11 trillion by 2030. This growth is fueled by the sharp decline in cost per mile expected with autonomous technology. According to ARK's analysis, the addressable market expands rapidly at lower price points:

- **$2-$4 per mile:** Represents today's traditional ride-hail services, with a modest market of $4 to $30 billion.

- **$1 per mile:** Opens up nearly $990 billion in new ridership, replacing commuting miles in Western markets.

- **$0.60 per mile:** Unlocks $2.4 trillion in non-commuting trips across higher-income regions.

- **$0.50 per mile:** Targets $2.75 trillion in demand, especially in lower-income markets.

- **$0.25 per mile:** Provides the most significant opportunity—$5 trillion—for low-cost, autonomous travel, driving 30 trillion addressable miles.

As autonomous mobility becomes more affordable, it will lower barriers for underserved regions and demographics, transforming urban and rural transit. Freight operations will also benefit, with lower labor and fuel costs increasing efficiency. By 2030, autonomous platforms are expected to dominate, enabling cost-effective, accessible transportation that brings us closer to an integrated, autonomous future.

National Highway Traffic Safety Administration 2029 Automatic Emergency Braking Systems Regulations and Impact on AV and Transportation

As outlined in the blog "Leading in Safety: How OEMs Can Meet NHTSA's Stringent AEB Requirements While Controlling Costs". (https://www.zendar.io/blog/meet-nhtsa-aeb-regulations-while-controlling-costs) in April 2024, the National Highway Traffic Safety Administration (NHTSA) introduced regulations requiring all new passenger vehicles and light trucks sold in the U.S. to include advanced automatic emergency braking (AEB) systems by 2029. These systems must function effectively at speeds up to 90 mph, perform reliably with low-beam headlights at night, and entirely prevent collisions at speeds up to 62 mph for vehicles and 45 mph for pedestrians.

Preliminary NHTSA tests revealed that only one of 13 vehicle models from 2023 met these standards, and even then, the braking could have been more harsh. Upgrades are needed for most vehicles to comply by 2029.

Implementing advanced AEB systems is expected to significantly impact manufacturing costs. Initial NHTSA estimates put industry-wide costs at $282 million annually for software updates. However, automakers argued that additional hardware, like enhanced sensors, would be required, leading to revised estimates of $354 million. Industry groups forecast even higher costs, projecting $430 million per manufacturer annually, potentially reaching billions.

Despite the costs, these regulations aim to save lives and reduce injuries. NHTSA estimates the standards could save 360 lives and prevent 24,000 injuries annually by reducing collisions with vehicles and pedestrians, including at night.

These AEB requirements are expected to accelerate autonomous vehicle (AV) development. AEB systems are key components of advanced driver assistance systems (ADAS), foundational for AVs. Standardizing these technologies will push automakers to invest in sophisticated sensors and processors, bringing the industry closer to fully autonomous vehicles.

While costly, the regulations drive innovation. Automakers are exploring solutions like AI-powered radar perception technologies to meet the standards without significant hardware upgrades. Companies, such as Zendar, offer software that enhances radar systems, enabling them to detect and classify objects efficiently and cost-effectively.

NHTSA's AEB mandate promises to improve road safety and support AV advancement. While compliance poses financial challenges, it fosters innovation, leading to safer, more innovative vehicles and a more advanced transportation ecosystem.

As discussed in the article, "Information of Waymo's L4 and Tesla's L2 AV Systems – How to Design a Safer Road System with AV Technology" both human drivers (especially new or drunk, drowsy, doing drugs or distracted driver human drivers) and AV systems need to be monitored to make sure they drive properly and within the limits of their ability.

The good news is: if the Waymo L4 and L2 Tesla systems are used correctly, you are much safer than in a car with a human driver without this technology. The bad news is if you use an L2 system as if it were an L5/L4, you are a much more dangerous driver. For L2 systems like Tesla FSD AI systems in 2024, the human driver is liable for the AI system. Waymo's L4 system takes the liability as discussed in the article, "Responsible AI - When the AI Has to Be Right: Who Holds the Reins? Unpacking Liability in Healthcare, Law, and Transportation", LinkedIn. It will be important to track when automakers take liability for their L3 and L4 solutions. When AI systems are tasked with making decisions that affect human lives, the need for accountability is not just important—it's imperative. The key question to ask is: who holds the reins when things go wrong? Is it the developers who created the AI, the professionals who use it, or the organizations that deploy these systems, or the driver?

The Future of Autonomous Trucking in 2030

The U.S. trucking industry is a cornerstone of the nation's economy, responsible for transporting approximately 72.6% of all freight by weight in 2022.

Technological Advancements: Autonomous trucks now operate with advanced sensor arrays, including LiDAR, radar, and high-definition cameras, enabling precise navigation and obstacle detection. Machine learning algorithms have evolved, allowing these vehicles to adapt to diverse road conditions and traffic scenarios. Continuous data collection and analysis have refined these systems, ensuring reliability and safety.

Economic Impact: Integrating autonomous trucks could address the longstanding driver shortage, reducing operational costs and increasing delivery efficiency. Companies have reallocated human resources to roles requiring complex decision-making and customer interaction. The industry has seen a surge in productivity, with trucks operating around the clock without the limitations of human drivers.

Regulatory and Safety Considerations: Comprehensive regulation is needed to establish and govern autonomous trucking, focusing on safety standards, cybersecurity, and environmental impact. Collaborations between industry leaders and policymakers will ensure these vehicles meet stringent safety criteria, gaining public trust and acceptance.

Environmental Benefits: Autonomous trucks can contribute to environmental sustainability by optimizing routes and driving behaviours, reducing fuel consumption and lowering emissions. The adoption of electric and alternative fuel-powered autonomous trucks can further minimized the carbon footprint of the logistics sector.

It may be at least 10 years before regulations allow a truck to drive without a human safety driver in the cab. As AV technology advances, truck drivers can engage in various tasks while the AV system handles driving, ensuring they are prepared to intervene if necessary. These activities enhance operational efficiency and comply with federal Hours of Service (HOS) regulations.

1. **Rest and Sleep:** Utilize onboard sleeper berths to rest, allowing drivers to extend their driving hours within HOS limits.

2. **Route Planning:** Review and adjust upcoming routes based on real-time traffic and weather data to optimize delivery schedules.

3. **Logbook Management:** Update electronic logbooks to ensure accurate tracking of driving and rest periods, maintaining compliance with HOS regulations.

4. **Vehicle Monitoring:** Oversee vehicle diagnostics and system alerts to promptly identify and address maintenance needs.

5. **Cargo Inspection:** Monitor cargo conditions using onboard sensors to ensure load security and integrity.

6. **Communication:** Stay in contact with dispatchers and clients to provide updates on delivery status and address any concerns.

7. **Regulatory Compliance:** Review and ensure adherence to transportation regulations, including weight limits and hazardous material handling.

8. **Training and Education:** Engage in ongoing training modules to stay updated on AV technology and safety protocols.

9. **Documentation:** Complete necessary paperwork, such as bills of lading and delivery receipts, to streamline administrative processes.

By integrating these tasks into their routine, drivers can enhance productivity and safety while remaining ready to take control of the AV system requires human intervention.

FUTURE OUTLOOK

The success of autonomous trucking has paved the way for advancements in other areas of transportation, including autonomous passenger vehicles and urban delivery systems. Ongoing research and development continue to enhance the capabilities of these systems, promising even greater efficiency and safety in the future. The autonomous trucking industry in 2030 will be a testament to technological innovation and strategic foresight.

How Transportation safety impacts healthcare and reaching zero transportation deaths.

In 2014, traffic safety trends were going in the right direction. Even though any avoidable road system death is one death too many, unfortunately, that changed starting in the last 10 years.

2016 Was the Deadliest Year on American Roads in Nearly a Decade:

In 2016, data shared by the National Safety Council estimates that as many as 40,000 people died in motor vehicle crashes, a 6% rise from 2015. This was a 14% increase in deaths since 2014, the latgest two-year jump in more than five decades. (http://fortune.com/2017/02/15/traffic-deadliest-year/)

Unfortunately, road system deaths have only gotten worse. At the end of 2023, the World Health Organization announced:

> "Every year, the lives of approximately 1.19 million people are cut short due to a road traffic crash. Between 20 and 50 million more people suffer non-fatal injuries, with many incurring a disability."

"Road traffic injuries", 2023 (https://www.who.int/news-room/fact-sheets/detail/road-traffic-injuries).

The National Safety Council report "Motor Vehicle – Introduction – Injury Facts", (https://injuryfacts.nsc.org/motor-vehicle/overview/introduction/) reported that in 2021, motor vehicle accidents resulted in 5.4 million injuries, costing the economy a staggering $498 billion. Tragically, 46,890 lives were lost to motor vehicle crashes, marking an upward trend in fatalities and injuries since 2014.

The escalation in road traffic deaths underscores an urgent need to prioritize a safer road system. With the historical context showing a significant reduction in deaths due to safety standards, the recent reversal in this trend is alarming. The rise in fatalities and injuries, alongside the economic and societal costs, highlights the road system as a significant public health issue. Focusing on healthcare must include efforts to enhance road safety, utilizing advancements in technology such as automated and autonomous vehicles, which hold the potential to dramatically reduce accidents. This approach promises to save lives and alleviate the economic burden on society, making it an indispensable facet of public health strategy moving forward.

As presented by Darrel M. West in "Moving forward: Self-driving vehicles in China, Europe, Japan, Korea, and the United States", in 2016, "The World Economic Forum estimated that the digital transformation of the automotive industry would generate $67 billion in value for that sector and $3.1 trillion in societal benefits. That includes improvements from autonomous vehicles, connected travellers, and the transportation enterprise ecosystem as a whole." (https://www.brookings.edu/wp-content/uploads/2016/09/driverless-cars-2.pdf)

As reported by NHTSA's "Traffic Crash Deaths: Early Estimates Jan-March 2024," from 2014 to 2021, there was an increase of 10,486 traffic deaths, representing a 32% increase.

Deaths have decreased by ~5% in the last 2 years, but each of those deaths is a tragedy that, in most cases, did not need to happen if the right technology was used properly.

The intersection of transportation safety and healthcare represents a critical frontier in public health and societal well-being. The staggering statistics of road fatalities and injuries not only highlight a pressing safety issue but also underscore a significant burden on healthcare systems worldwide. As we've seen, the economic cost of road accidents runs into hundreds of billions of dollars annually, diverting crucial resources from other healthcare and social development areas.

Advanced automated and autonomous vehicle technologies present a promising path toward this zero-fatality goal. By reducing human error, which accounts for the vast majority of road accidents, these technologies have the potential to save over a million lives. However, their successful implementation requires a coordinated effort involving automakers, tech companies, policymakers, and the public.

The healthcare implications of improved transportation safety extend beyond reduced emergency room visits and lower mortality rates. A safer transportation system could lead to decreased long-term care needs for accident survivors, reduced strain on mental health services for those affected by accidents, and overall improved quality of life for communities. Moreover, the resources currently allocated to treating accident victims could be redirected to other pressing healthcare needs, potentially transforming the public health landscape.

However, as we transition to using new transportation technologies, we must remain vigilant about potential new risks and challenges. The introduction of autonomous vehicles, for instance, may bring unforeseen safety issues that will require ongoing research, regulation, and public education.

By keeping safety at the forefront of this transportation revolution, we can create a future where road fatalities are a rarity rather than a daily occurrence, profoundly impacting public health and quality of life for generations.

The path to zero transportation deaths is challenging. Still, the potential rewards for saved lives, optimized healthcare resources, and improved communities make it a goal worth pursuing with utmost dedication and innovation.

What is CASE: Connected, Automated, Shared, Electric?

The automotive industry is undergoing a profound transformation, driven by four key technological trends known collectively as CASE: Connected, Automated, Shared, and Electric. This acronym encapsulates the significant forces reshaping vehicles and the entire concept of mobility in the 21st century.

Connected: Connectivity in vehicles refers to integrating internet and network-based services into transportation. Modern cars are increasingly becoming "computers on wheels," with advanced infotainment systems, real-time navigation, and vehicle-to-everything (V2X) communication capabilities. This connectivity enables vehicles to interact with their environment, other vehicles, and infrastructure, enhancing safety, efficiency, and user experience. For instance, connected cars can receive real-time traffic updates, automatically report accidents and schedule maintenance based on continuous system monitoring.

Automated: Vehicle automation ranges from essential driver assistance features to fully autonomous driving capabilities. As noted earlier, the Society of Automotive Engineers (SAE) defines six levels of automation, from Level 0 (no automation) to Level 5 (full automation). As of 2024, most consumer vehicles operate at Levels 1 or 2, with some high-end models achieving Level 3 in specific conditions. The push towards higher levels of automation aims to dramatically reduce accidents caused by human error, improve traffic flow, and provide mobility options for those unable to drive conventional vehicles.

Shared: Shared mobility represents a shift away from traditional car ownership models. This encompasses ride-sharing services, car-sharing platforms, and other innovative mobility solutions that optimize vehicle usage and reduce the need for individual car ownership. The rise of companies like Uber, Lyft, and various car-sharing startups exemplifies this trend. As autonomous technology advances, the potential for shared autonomous vehicles could further revolutionize urban transportation, reducing the number of cars on the road and increasing accessibility.

Electric: The transition to electric vehicles (EVs) is the most visible aspect of the CASE revolution. Driven by environmental concerns and advancements in battery technology, the automotive industry is rapidly shifting towards electrification. EVs offer numerous benefits, including zero tailpipe emissions, lower operating costs, and reduced dependence on fossil fuels. The growth of the EV market is accompanied by the expansion of charging infrastructure and innovations in battery technology, aiming to address range anxiety and charging time concerns.

The Interplay of CASE Technologies: While each element of CASE is significant, its revolutionary potential lies in its integration. For example, connected and automated technologies can optimize traffic flow and reduce congestion. Shared electric vehicles could provide efficient, environmentally friendly urban mobility solutions. Automated electric vehicles could form the backbone of new public transportation systems or revolutionize long-haul freight transport.

Challenges and Opportunities: Implementing CASE technologies presents challenges and opportunities. Issues such as data privacy and security, regulatory frameworks for autonomous vehicles, and the need for substantial infrastructure investments must be addressed. However, the potential benefits are immense, including improved road safety, reduced environmental impact, enhanced mobility for all segments of society, and new economic opportunities in the technology and service sectors.

Looking Ahead: As we approach 2030, the convergence of CASE technologies is expected to accelerate, fundamentally altering the automotive landscape. Cities may need to reimagine urban planning to accommodate these new mobility paradigms. The traditional automotive industry will likely see further disruption, with new players entering the market and established companies adapting to remain competitive.

The CASE revolution represents not just a transformation of vehicles but a reimagining of mobility itself. As these technologies mature and integrate, they have the potential to create transportation systems that are safer, more efficient, more accessible, and more sustainable than ever before, profoundly impacting how we live, work, and interact with our environment.

The Safest and Greenest Mile You Can Travel is the One You Do Virtually.

6G Virtual Transportation could enable immersive, real-time experiences replicating in-person interactions' sensory and social aspects. Business meetings, educational experiences, and even tourism could be conducted virtually, drastically reducing the need for physical travel. This shift would not only save time and increase productivity but also lead to substantial reductions in transportation-related emissions.

Moreover, integrating 6G technology with smart city infrastructure and autonomous vehicles could optimize the remaining physical transportation needs. Real-time data analysis and AI-driven traffic management could minimize congestion and maximize efficiency, reducing emissions and energy consumption.

However, realizing this potential depends on responsible implementation and policy decisions. It's crucial to ensure that the energy demands of 6G networks and virtual reality systems don't offset the emissions savings from reduced physical travel.

As we stand at this technological crossroads, our choices will determine whether we harness the full potential of 6G Virtual Transportation to create a more sustainable, efficient, and connected world. By prioritizing environmental considerations in developing and deploying these technologies, we have the opportunity to revolutionize how we move and interact with our world, potentially ushering in a new era of sustainable living and dramatically reduced carbon emissions.

Will Self-Driving Arrive by 2030?

As we approach 2030, whether fully self-driving vehicles will become a widespread reality remains a complex and multifaceted question. While significant progress has been made in autonomous vehicle technology, the path to ubiquitous self-driving cars is more complicated than many predicted a decade ago.

By 2030, there will be an increase in vehicles with advanced driver assistance systems (ADAS) and conditional automation (SAE Level 3). The jump from Level

3 to fully autonomous vehicles (Levels 4 and 5) is more than just a technological leap—it involves overcoming regulatory, ethical, and societal challenges. While we expect to see growth in robotaxi services, with potentially millions of weekly rides by 2030, this represents just a fraction of overall transportation.

The impact of even this level of automation could be profound. We anticipate a 90% decrease in traffic accidents among vehicles equipped with Level 3 technology, leading to significantly lower insurance premiums for these vehicles. This progress in safety alone justifies the continued investment and development in autonomous technology.

Looking further ahead to 2040, our projections become more optimistic but still realistically tempered. We expect 90% of all vehicles to support automated Level 3 or more excellent driving, with drivers engaging it for 90% of their commute time. This could save commuters nearly an hour each day. Moreover, we anticipate that 20% of new vehicles sold will be SAE Levels 4 or 5, allowing for almost full automation during commutes and daily travel.

The potential benefits are enormous. By 2040, we could see a 50% reduction in all traffic accidents, dramatically lowering fatalities and insurance costs. The robotaxi industry could explode, accounting for 50% of all global ridesharing trips.

However, it's crucial to remember that these projections are not guarantees. The development of self-driving technology has faced unforeseen challenges, from technical hurdles to public acceptance issues. The ethical implications of autonomous decision-making in critical situations remain a topic of intense debate. Moreover, the regulatory landscape will need to evolve significantly to accommodate fully autonomous vehicles on a large scale.

The environmental impact of this shift is also a critical consideration. While the potential for reducing emissions through more efficient driving and reduced congestion is significant, the increased accessibility and convenience of autonomous vehicles could also lead to more miles travelled overall. Pairing the rollout of autonomous technology with sustainable energy solutions will be crucial to ensure a net positive environmental impact.

While fully self-driving vehicles may not be ubiquitous by 2030, we are undoubtedly on the path to a transportation revolution. The next decade will likely see a dramatic increase in partially automated vehicles, setting the stage for more advanced autonomous systems in the following years. The key to realizing the full potential of this technology lies in continued research and development, thoughtful regulation, and public education.

As we move forward, it's essential to approach self-driving technology's development with optimism and caution. The potential benefits of safety, efficiency, and accessibility are immense, but so are the challenges we must overcome. By 2030, we may not have fully achieved the fully autonomous future once envisioned, but we will undoubtedly be well on our way to transforming transportation as we know it. The journey towards self-driving vehicles is not just about technological advancement—it's about reimagining our relationship with transportation and shaping a safer, more efficient, and more sustainable future for all.

AUTONOMOUS MOBILITY IN 2030

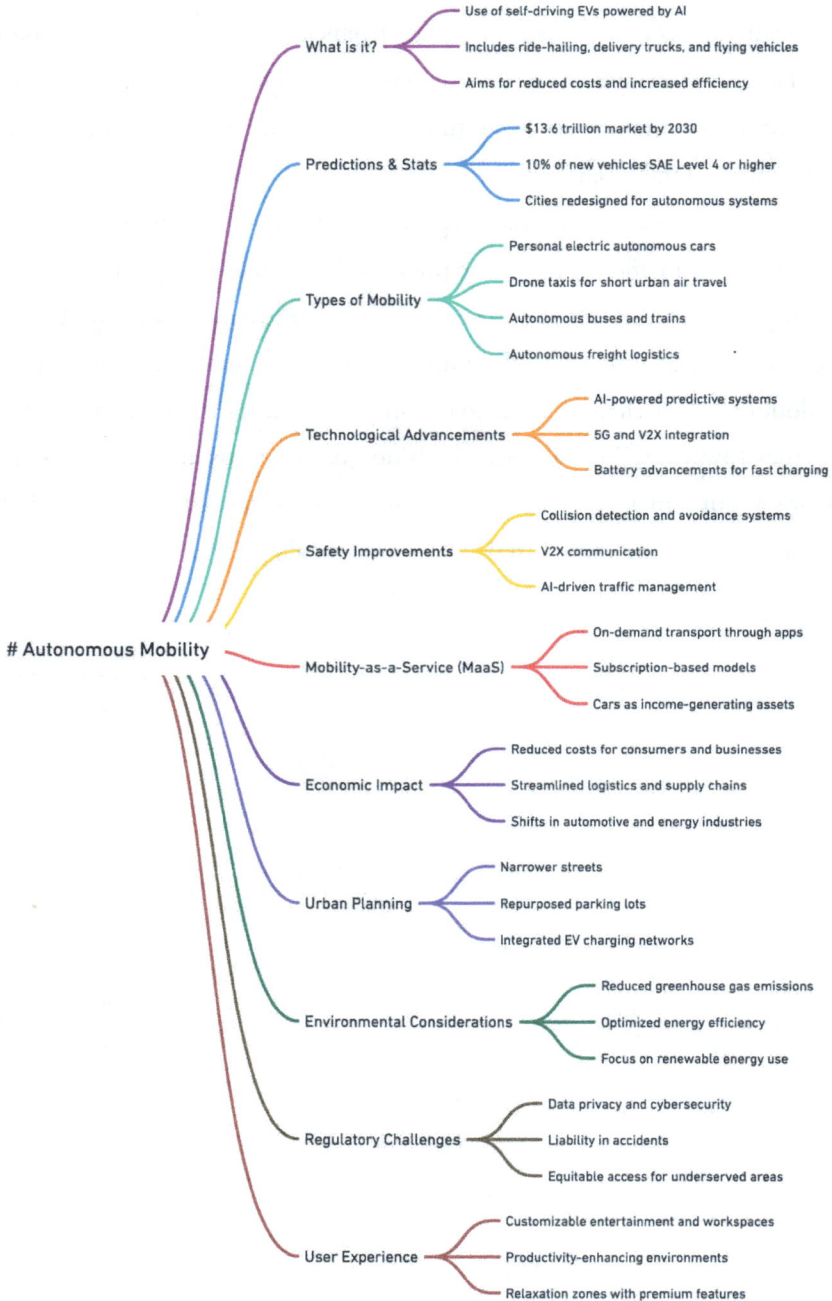

Autonomous Mobility

- **What is it?**
  - Use of self-driving EVs powered by AI
  - Includes ride-hailing, delivery trucks, and flying vehicles
  - Aims for reduced costs and increased efficiency

- **Predictions & Stats**
  - $13.6 trillion market by 2030
  - 10% of new vehicles SAE Level 4 or higher
  - Cities redesigned for autonomous systems

- **Types of Mobility**
  - Personal electric autonomous cars
  - Drone taxis for short urban air travel
  - Autonomous buses and trains
  - Autonomous freight logistics

- **Technological Advancements**
  - AI-powered predictive systems
  - 5G and V2X integration
  - Battery advancements for fast charging

- **Safety Improvements**
  - Collision detection and avoidance systems
  - V2X communication
  - AI-driven traffic management

- **Mobility-as-a-Service (MaaS)**
  - On-demand transport through apps
  - Subscription-based models
  - Cars as income-generating assets

- **Economic Impact**
  - Reduced costs for consumers and businesses
  - Streamlined logistics and supply chains
  - Shifts in automotive and energy industries

- **Urban Planning**
  - Narrower streets
  - Repurposed parking lots
  - Integrated EV charging networks

- **Environmental Considerations**
  - Reduced greenhouse gas emissions
  - Optimized energy efficiency
  - Focus on renewable energy use

- **Regulatory Challenges**
  - Data privacy and cybersecurity
  - Liability in accidents
  - Equitable access for underserved areas

- **User Experience**
  - Customizable entertainment and workspaces
  - Productivity-enhancing environments
  - Relaxation zones with premium features

KEY PREDICTIONS & STATISTICS

- The global autonomous vehicle market is expected to reach $13.6 trillion by 2030

- ~10% of new vehicles sold globally are projected to be at level 4 or higher autonomy

- By 2030, autonomous cars are expected to dominate both passenger transport and freight logistics

- Drone taxis will enable short-distance air travel in urban centers by 2030

- Urban planning will shift dramatically by 2030, with cities redesigned for autonomous electric vehicles, public transit, and micro-mobility. Streets will be narrower, and parking lots will be repurposed for green spaces or housing

WHAT IS IT?

Autonomous mobility refers to using self-driving electric vehicles (EVs) powered by artificial intelligence (AI) and advanced battery technology to transport people and goods without human intervention. These autonomous systems span various mobility forms, including ride-hailing services, delivery trucks, freight logistics, and flying vehicles. This shift from human-driven cars to fully autonomous platforms is expected to change transportation infrastructure, reducing costs and drastically increasing efficiency.

WHAT IS GOING TO HAPPEN?

By 2030, autonomous vehicles (AVs) will dominate passenger transport and freight logistics. The cost per mile for transportation in 2022 was $0.72 per mile. This will decrease dramatically, potentially falling to $0.25 per mile, unlocking trillions of dollars in new demand for mobility services. The ride-hailing market alone could expand from $300 billion today to over $11 trillion by 2030. Autonomous electric vehicles will offer cheaper, more efficient transportation, drive down freight labour and fuel costs, and provide accessible, low-cost travel options for underserved communities. Cities and infrastructure will evolve to accommodate these vehicles, and rural areas will see enhanced connectivity.

WHY DOES IT MATTER?

This revolution in transportation will have far-reaching economic and social implications. Autonomous vehicles will reduce travel costs, making transportation more affordable and scalable. This will democratize access to mobility for low-income regions, improve public transit efficiency, and reduce carbon emissions through electrification. For businesses, autonomous freight operations will streamline logistics and supply chains, cutting down delivery times and operating costs. On a broader scale, it will reshape urban planning, reduce traffic congestion, and potentially lower accident rates by removing human error.

WHO WILL IT IMPACT?

The rise of autonomous mobility will impact multiple industries: transportation, logistics, automotive manufacturing, energy, and urban infrastructure. It will affect drivers as demand for human-operated vehicles will decline, leading to shifts in employment. Consumers will benefit from cheaper and more accessible mobility options, while businesses will see lower transportation costs. Governments must adapt regulations, public transportation systems, and infrastructure to accommodate these autonomous vehicles. Freight companies will also see significant cost reductions and efficiency improvements, reshaping global trade and logistics.

CALL TO ACTION

Governments, industries, and individuals must invest in autonomous mobility infrastructure, including smart roads, EV charging networks, and AI-powered traffic management systems, to prepare for this transformation. Policymakers should create regulatory frameworks encouraging the safe and widespread adoption of autonomous vehicles while focusing on reskilling workers displaced by automation in the transport sector. Preparing now will ensure a smooth transition to an autonomous mobility future by 2030.

Pro Tip: If you are in a region where autonomous vehicles are already on the road (San Francisco, CA, Austin, TX, Beijing, CN), try it. Don't wait until all cars are autonomous; take a ride in one while they are still notable and not part of everyday life.

TYPES OF MOBILITY IN 2030

By 2030, mobility will encompass a broad range of advanced transportation solutions. Personal vehicles will primarily be electric and autonomous, with many opting for mobility-as-a-service (MaaS) over car ownership. Taxi services will transition to autonomous fleets, making rides cheaper and more efficient. Mass transportation will include self-driving buses, electric shuttles, and high-speed Maglev trains, offering rapid inter-city travel. Drone taxis will enable short-distance air travel in urban centers, while shipping routes will be dominated by autonomous electric and wind-powered vessels for freight and passenger travel. Spacecraft for commercial space travel will become more accessible, and airplanes will integrate sustainable fuels and autonomous piloting systems for safer, more efficient flights.

Technological Advancements

By 2030, transportation will have evolved into a high-tech ecosystem, blending AI, machine learning, and advanced battery technologies to power autonomous systems across personal, public, and commercial vehicles. Self-driving vehicles, now standard, use AI not only for navigation but also to predict and adapt to traffic patterns, reducing accidents and improving safety. Battery advancements will enable ultra-fast charging, with EVs capable of charging fully in minutes and offering significantly longer ranges, making them more practical for long-distance travel.

Hyperloop and Maglev trains will redefine city-to-city travel, providing ultra-fast, low-energy transportation. These systems will operate autonomously and integrate with AI-powered scheduling systems to optimize travel flow, ensuring rapid transit with minimal energy consumption. Drone technology will extend beyond delivery, with passenger and cargo drones offering quick, on-demand aerial travel. Advanced swarm technology will allow drones to coordinate efficiently in shared airspace, transporting people and goods safely above urban congestion.

Integration with 5G and 6G networks will support real-time vehicle-to-everything (V2X) communication, allowing vehicles to interact with each other and with surrounding infrastructure like traffic signals, road sensors, and pedestrian devices. This interconnected system will enable adaptive traffic flow, reduce congestion, and improve urban mobility. By leveraging these advancements, autonomous mobility in 2030 will be faster, greener, and safer than ever, with seamless transitions between multiple modes of travel.

Safety Improvements

By 2030, safety in autonomous mobility will be significantly enhanced through a fully integrated network of AI-driven sensors, predictive algorithms, and real-time data analytics across all types of vehicles and infrastructure. Collision detection systems will use multiple sensory inputs—such as lidar, radar, and cameras—working harmoniously with AI algorithms to detect and respond to obstacles in milliseconds, reducing accidents due to delayed human reaction times. Predictive AI will analyze traffic patterns, pedestrian movement, and environmental conditions to anticipate potential hazards, allowing autonomous vehicles to make preemptive adjustments in real time.

Vehicle-to-everything (V2X) communication will extend beyond vehicle-to-vehicle interaction to include vehicle-to-infrastructure (V2I) and vehicle-to-pedestrian (V2P) communication. Vehicles will interact seamlessly with traffic lights, road signs, and crosswalks, enabling dynamic traffic management and adaptive speed adjustments, which optimize traffic flow while reducing collisions. Autonomous traffic systems in densely populated urban centers will control traffic lights, adjust speed limits, and manage lane allocations dynamically to prevent bottlenecks and streamline mobility.

In the air, autonomous flight paths for drones and passenger air taxis will be governed by AI-powered air traffic control systems capable of coordinating thousands of simultaneous flights. These systems will manage dedicated drone corridors, preventing collisions, and ensuring safe navigation above urban areas. For emergencies, drone-based rescue and first aid services will reach accident sites within moments, delivering critical assistance before human responders arrive.

Further advancements in AI autopilots for conventional and vertical takeoff aircraft will also improve aviation safety, allowing for automated rerouting in response to weather changes, air traffic, and other conditions. These autopilot systems will work alongside health monitoring technologies within autonomous vehicles, tracking passengers' vital signs and detecting abnormalities, alerting emergency services if needed.

The reduction in human involvement and the complete integration of AI safety systems, real-time data, and interconnected infrastructure will likely result in a drastic reduction in accidents across all forms of transportation, bringing us closer to a world with near-zero transportation-related fatalities.

Mobility-as-a-Service Revolution

By 2030, the Mobility-as-a-Service (MaaS) revolution will redefine personal and shared transportation. People can book any transportation they need through a single, user-friendly platform—autonomous cars, electric scooters, passenger drones, or high-speed Maglev trains. MaaS will generate personalized travel plans optimized for time, cost, and environmental impact, allowing users to seamlessly shift between different transportation modes without delays or additional charges. With the rise of affordable MaaS subscription models, car ownership will no longer be necessary. Instead, most people will access various transport options on demand, reducing personal transportation costs and traffic congestion. In cities, shared and public transportation will become the primary modes of travel, democratizing access and promoting environmental sustainability.

One of the most exciting aspects of the MaaS revolution is the way it transforms personal vehicles into income-generating assets. Autonomous cars can operate independently, turning downtime into profitable hours. For example, a personal car could be out working as a rideshare or delivery service while its owner is asleep or simply not needing the car. Car owners can rent out their vehicles during idle hours, making money passively without being directly involved. In effect, vehicles will become revenue-generating tools rather than idle, depreciating assets parked for most of the day.

This new value proposition opens up income-generating opportunities for car owners and automakers. Car manufacturers will likely offer programs that streamline renting, maintenance, and insurance for autonomous vehicles in shared fleets, allowing them to collect a share of rental income. Subscription and leasing models may include built-in rental options, enabling car owners to let their vehicles join autonomous MaaS fleets for extra income when they aren't in use. As a result, automakers can diversify revenue streams, moving beyond one-time car sales to ongoing profit from shared mobility programs.

MaaS presents a new form of economic empowerment for consumers: vehicles become a source of passive income, available to anyone who wants to generate extra revenue. MaaS will mean reduced car ownership, lower emissions, and less congestion for cities and communities as people rely on shared and public autonomous transport. By 2030, MaaS will transform the entire concept of car ownership, making transportation not only more accessible but also a source of additional income and value for all.

Economic Impact

The economic impact of autonomous and electric mobility will be transformative, affecting both global industries and individual consumers alike. With autonomous systems driving down the cost per mile for travel, operational expenses for businesses will plummet, making transport cheaper and more accessible. Consumers will feel this shift, too, with lower transportation costs, making daily commutes, long-distance travel, and goods delivery more affordable. The ride-hailing industry is set to expand exponentially, projected to grow from $300 billion to over $11 trillion by 2030, as the convenience and efficiency of autonomous, shared transportation make it the default choice for many.

Autonomous freight transport will streamline global supply chains, cutting labour and fuel costs and increasing efficiency. Autonomous fleets can operate around the clock, significantly reducing delivery times and enabling "just-in-time" supply chain management, which is crucial for many industries. This efficiency will lower shipping costs across sectors, benefiting companies and consumers.

Additionally, the widespread adoption of electric vehicles (EVs) and alternative fuel vehicles (i.e. hydrogen, ethanol) will reduce reliance on fossil fuels, cutting fuel costs and shifting economic dependence away from oil-producing countries and companies. As EVs become mainstream, entire industries will face disruption: the insurance sector must adjust as accident rates drop with autonomous driving. In contrast, traditional car manufacturing must evolve as demand shifts towards autonomous and electric models. These economic shifts will ultimately drive a major restructuring across sectors, creating new opportunities and challenges in a rapidly changing global economy.

Societal Changes

Autonomous mobility will bring transformative societal changes, particularly for underserved groups like the elderly, disabled, and young people. These groups will have greater independence, as autonomous vehicles will offer accessible, on-demand transportation without the need for driving skills. This increased mobility will enhance their ability to participate in social, educational, and economic activities, leading to a more inclusive society. Additionally, reducing car ownership and reliance on public transportation will create more equitable access to mobility services for low-income communities.

Urban Planning and Infrastructure

Urban planning will be reimagined by 2030 as cities adapt to autonomous electric vehicles (EVs), public transit innovations, and micro-mobility solutions like e-scooters and shared bikes. With fewer people owning personal cars, traditional road infrastructure will undergo significant changes: streets will be narrower and optimized for pedestrians and light electric vehicles, while parking lots and garages will be repurposed for green spaces, community hubs, or residential housing. This shift will reclaim urban space, making neighbourhoods more walkable and improving cities' aesthetic and environmental quality.

To support the growing network of autonomous EVs, charging stations will be strategically integrated into existing infrastructure. At the same time, smart roads equipped with sensors and real-time data communication systems will guide autonomous cars and ensure traffic flows smoothly. Cities will also establish dedicated lanes for autonomous buses and flying drone taxis, creating a seamless, layered transport network that efficiently handles different transit modes. This new urban design, paired with cleaner electric mobility, will drastically reduce congestion and lower air pollution levels, contributing to an enhanced quality of life in urban areas where noise, traffic, and pollution once dominated.

Environmental Considerations

The shift to electric and autonomous mobility will bring extensive environmental benefits, marking a significant step toward sustainable urban ecosystems. Electric vehicles (EVs) powered by renewable energy sources—such as nuclear, solar, wind, and, eventually, hydrogen—will drastically reduce greenhouse gas emissions, making transportation far clearer than gasoline-powered vehicles. This transition will help mitigate air pollution in densely populated cities and contribute to the decarbonization goals set by global climate agreements. With EV adoption, the autonomous systems embedded in these vehicles will further optimize energy efficiency by selecting the most fuel-efficient routes, reducing unnecessary idling, and managing speed to minimize energy use, thereby shrinking the overall environmental footprint of the mobility sector.

Using lightweight materials in vehicle manufacturing and advancements in battery recycling will reduce resource consumption. By 2030, the mobility sector will be critical in the global effort to combat climate change.

Regulatory Challenges

As mobility technologies advance, regulation will become a critical challenge for governments worldwide, requiring comprehensive frameworks that balance innovation, safety, and equity. Autonomous vehicle safety standards must be redefined, ensuring that self-driving vehicles meet rigorous testing and performance criteria before they are allowed on public roads. As these vehicles rely on vast amounts of data, governments must establish stringent data privacy protocols to protect users' personal information and prevent potential misuse by third parties or cybersecurity threats. Defining liability in autonomous vehicle accidents presents another major hurdle: regulators must clarify who is responsible—whether the manufacturer, software developer or user—when an AI-driven vehicle is involved in an accident.

Drone taxis and passenger drones will further complicate regulation as they take to the skies, requiring new standards for airspace management to prevent collisions and ensure safe routes within densely populated urban areas. These drones must comply with designated flight paths, speed limits, and safety protocols, creating the need for AI-driven air traffic control systems or "drone corridors" to effectively manage urban airspace. This regulatory framework will involve collaboration across sectors, from aviation authorities to tech firms, ensuring that airspace is safely shared between traditional aircraft, drones, and other airborne mobility solutions.

The shift from human-driven vehicles to autonomous systems will also trigger ethical and legal questions about job displacement, as automation reduces the need for human drivers in industries like trucking, logistics, and ride-hailing. Governments must balance support for these affected workers through job retraining programs or economic assistance while enabling the continued progress of mobility innovations. As autonomous technologies create new forms of employment and monetary value, regulators will ensure the transition is socially responsible.

Equitable access to autonomous mobility will be essential to avoid creating mobility "haves and have-nots." Regulations must ensure that autonomous transport is available in underserved areas, not just major urban centers, enabling broader access for all demographics. Balancing the push for innovation with public safety, privacy, and access will require a proactive approach from policymakers to make autonomous mobility safe, fair, and beneficial for society.

User Experience and Productivity

By 2030, the user experience in mobility will be centred on convenience, personalization, and productivity. Autonomous vehicles will provide custom entertainment, workspaces, or relaxation environments, turning travel time into productive or leisure time. With no need to focus on the road, passengers can fully engage in various activities, and vehicle interiors will be designed to maximize comfort and versatility. Imagine passenger seats that swivel to face each other, creating a lounge-like setup perfect for socializing, collaborative work sessions, or family time.

For those looking to unwind, autonomous vehicles will offer built-in entertainment systems that rival home theatres: large screens, premium surround sound systems, and personalized streaming options. Riders can binge-watch their favourite shows, listen to music through immersive soundscapes, or gamble on high-definition screens. For more interactive experiences, passengers can strap on a VR headset, transforming the vehicle interior into a virtual adventure space—whether it's a serene escape into nature or an interactive video game. With augmented reality (AR), passengers could experience a virtual tour synced to the passing scenery, allowing travellers to learn about historical landmarks or local highlights as they drive by.

Productivity options will be abundant for those looking to work. Vehicles can quickly transform into mobile workspaces with high-speed internet, charging stations, and privacy settings. Adjustable seating, noise-cancelling environments, and virtual meeting capabilities will allow professionals to hold video conferences, review documents on large shared screens, or brainstorm with colleagues across virtual whiteboards.

Personalized ambient lighting, climate control, and aromatherapy systems will create a spa-like atmosphere for those who view the vehicle as a relaxation zone. Reclining seats with massage functions, calming audio, or guided meditation will transform a simple commute into a moment of zen, ideal for unwinding after a long day.

Market Projections and Adoption Rates

By 2030, autonomous electric vehicles are expected to dominate the mobility market, with high adoption rates in both personal and public transportation.

Ride-hailing services powered by autonomous platforms will grow to $11 trillion in value, while electric vehicles will make up most new car sales. Autonomous freight and drone delivery systems will revolutionize logistics, becoming mainstream in most industries. Governments and enterprises will continue investing in charging infrastructure, smart cities, and AI-driven transportation systems, ensuring rapid market penetration and widespread adoption.

AI, AGI, ASI, and Quantum Computing

On a mild April day in the year 2030, Samantha woke up in her sleek New York apartment to the gentle voice of Qubit, her personalized AI assistant powered by a decentralized cloud quantum computer. "Good morning, Samantha. Based on your restless night, I recommend a light breakfast with extra protein for energy." As an innovative corporate lawyer working for a hedge fund specializing in AI-driven investments, Samantha relied on Qubit to manage everything from her health to her work schedule accurately. It would be an exciting day—Samantha will be finalizing an investment in a one-person startup using autonomous agent swarms to solve global logistics challenges. By two months in, the startup was making $250K monthly revenue. What once required entire teams can now be done by an individual leveraging the vast computing power of AI. Many such startups are making millions in monthly revenue, a new normal in the AI economy.

Meanwhile, Samantha's partner, Othello, began his morning differently. As a celebrated author, he still writes the "old-fashioned" way—without AI. This human touch has become a rare, coveted art form in a world where Generative AI produces books and media on demand, providing endless variety. To protect the authenticity of his work, a live blockchain-traced camera feed tracked his every keystroke, proving the originality of each sentence he typed. This blockchain system allowed Othello to sell his works globally, instantly reaching readers without intermediaries, ensuring that his creative process and ownership remain undisputed. Deepfakes have become so realistic and pervasive that new laws have penalized those who fail to label AI-generated content.

Samantha went to her automated office in Midtown, where her hedge fund uses quantum encryption to securely analyze vast datasets in real time, predicting market trends faster than any traditional computing system could. The blockchain-backed AI systems allow the firm to invest seamlessly across global markets without intermediaries, moving billions of dollars securely and efficiently instantly. For Samantha, the power of AI and quantum computing has redefined how hedge funds operate, democratizing finance and providing unparalleled access to new markets. Micro investments, the latest up-and-coming asset class, have provided the much-needed wealth distribution that technology always promised but has just now been delivered.

Later in the day, Samantha met with her physician, Dr. Patel, who used quantum edge computers to analyze her genetic data from the analysis of her real-time urine stream, and one eyelash is carefully laid on the sensor. Dr. Patel created a personalized health plan using this data, including changing Samantha's physical routine and diet. This is synchronized with Qubit, which automatically ordered the new food options directly from her condo. Quantum technology has revolutionized healthcare, allowing doctors to identify health risks and create individualized treatment strategies that prevent issues before they arise.

In the evening, the couple met at a local art gallery where Othello showcased his latest work. Despite AI being a million times more powerful than the GPT-4 of 2023, Othello's human-authored works still captivate audiences, offering a unique perspective in a world increasingly dominated by algorithms. They reflected on how much their lives have changed in the last decade, where once-complicated processes like starting a company, investing globally, and selling art are now frictionless and secure, thanks to the integration of AI, AGI, ASI, and quantum computing into everyday life.

The future is here, and for Samantha and Othello, it's filled with opportunities previously unimaginable—a seamless blend of human creativity and machine power, driving both personal success and global transformation.

AI, AGI, ASI, and QUANTUM COMPUTING IN 2030

A mind map titled "# AI, AGI, ASI & Quantum Computing in 2030" with the following branches:

Key Predictions & Stats
- AI to contribute $15.7 trillion to global economy by 2030
- Personal AI assistants like "Qubit" to manage daily tasks
- Quantum sensors reduce waste in multiple industries

Technologies
- Artificial Intelligence (AI)
  - Automates tasks, optimizes decisions
  - Integrated in industries like healthcare, finance, and logistics
- Artificial General Intelligence (AGI)
  - Human-like reasoning across diverse tasks
  - Raises ethical and societal implications
- Artificial Super Intelligence (ASI)
  - Surpasses human intelligence in all domains
  - Poses significant ethical and governance challenges
- Quantum Computing
  - Exponentially faster calculations
  - Breakthroughs in cryptography, materials science, and AI

What is Going to Happen?
- AI will automate industries, improve efficiency, and transform professions
- Quantum computing operational maturity
- AGI approaching human-like intelligence

Why Does It Matter?
- AI and quantum computing redefine society
- Impacts on job markets, global security, and ethical decision-making
- Advances in health, logistics, and finance

Challenges and Concerns
- Job displacement due to automation
- Ethical dilemmas with AI and ASI
- Need for quantum-secure encryption to safeguard data

Economic and Societal Impact
- Cheaper, efficient operations in industries
- Changes in urban planning due to automation
- Redefined human labor with emphasis on creativity

Future of Work
- Shift towards AI collaboration roles
- New industries around human-AI interaction
- Universal Basic Income (UBI) may gain traction

Call to Action
- Invest in AI/quantum literacy and regulation
- Create ethical and equitable frameworks

Quantum Computing Milestones
- 1 million qubits projected by 2030
- Quantum error correction for reliable performance

Ethical Frameworks
- Fairness and transparency in AI systems
- Governance for ASI development

Recommended Reading
- "The Coming Wave" by Mustapha Suleyman
- "Superintelligence" by Nick Bostrom
- "Quantum Computing for Everyone" by Chris Bernhardt

KEY PREDICTIONS & STATISTICS

AI could contribute up to $15.7 trillion to the global economy by 2030:

- Terms like "agents" and "AGI" (Artificial General Intelligence) may fall out of use by 2030 as agentic behavior becomes a fundamental part of advanced AI systems, and the concept of AGI becomes less relevant in the face of highly specialized AI capabilities.

- By 2030, each person will have their own personal AI assistant, like Qubit, who manages many things we take time to do now. For example, saying 'Qubit, take me home' will summon an autonomous vehicle that will take you home directly without pulling out a device.

- Apps will be made 'as-required' and will be something you use and then they will disappear afterward.

- Quantum sensors will dramatically reduce waste in food, biotechnology, manufacturing, and construction by catching imperfections at the molecular level.

WHAT IS IT?

Artificial Intelligence (AI) refers to the development of machines capable of performing tasks that require human-like intelligence, such as pattern recognition, problem-solving, and decision-making. Artificial General Intelligence (AGI) takes this a step further by aiming to develop machines with human-level reasoning and learning abilities across various tasks. Artificial Superintelligence (ASI) goes beyond AGI, where machines surpass human intelligence in all domains. Quantum Computing harnesses the principles of quantum mechanics to process information exponentially faster than classical computers, particularly for solving complex problems like cryptography, material science, and AI training.

WHAT IS GOING TO HAPPEN?

By 2030, AI will continue to permeate every industry, improving predictions, efficiency, decision-making, and automation. Although AGI is still in development, there are some who regard it as already in use. It will be approaching human-like levels of problem-solving and learning across diverse tasks using a combination of autonomous agents. We will see early forms of Artificial Superintelligence (ASI) or strong AGI being explored, raising profound ethical and societal implications. Quantum computing will reach operational maturity with quantum machines performing calculations that classical computers cannot, enabling significant breakthroughs in fields like AI, materials science, drug discovery, and cybersecurity. Quantum computing, combined with AI, will supercharge advancements across disciplines, accelerating technological innovation. Until true billion-qubit quantum computers have arrived, neuromorphic quantum and quantum simulators will be a helpful resource that provides utility without potent, error-resistant quantum computers.

WHY DOES IT MATTER?

These technologies will redefine every aspect of society, from work and healthcare to governance and education. AI and AGI will improve industries by automating repetitive tasks, optimizing resource allocation, and enhancing problem-solving. ASI represents both enormous potential and existential risk: it could lead to unprecedented innovation or create challenges in governance, security, and ethical decision-making.

If algorithms are one million times smarter than an average human, which jobs will they render obsolete? McKinsey & Co. estimates that 400-800 million jobs will be displaced by automation by the year 2030. This and other ethical concerns surround this new superpower that will change the fabric of society in the blink of an eye.

Quantum computing will revolutionize industries that rely on complex computations like finance, logistics, and drug discovery, solving problems currently intractable to classical computing. If quantum computers achieve stable, practical performance, nearly all encryption systems—those safeguarding governments, banks, healthcare systems, defence, and transportation—will be rendered obsolete unless they are quantum-secure. In an instant, these quantum-capable systems could break through the encryption that forms the backbone of global digital security, making quantum-encrypted solutions essential to protect critical infrastructures and sensitive data worldwide.

WHO WILL IT IMPACT?

The rise of AI, AGI, ASI, and quantum computing will impact every sector of society. Industries such as healthcare, finance, education, and transportation will be transformed at unimaginable levels through automation, decision-making, and data processing. Governments will face new challenges related to regulation, security, and employment shifts. Individuals will experience profound changes in the labor market as AI systems take over many tasks, requiring new skills and training to adapt to an AI-driven economy. From daily interactions with your personal AI assistant to using swarms of agent AI to build entire organizations, no facet of society will be impacted by this new technological revolution by 2030. Buckle up!

CALL TO ACTION

Governments, businesses, and educational institutions must invest in AI and quantum literacy to prepare for the future. This includes updating regulatory frameworks to manage the risks of AI and quantum computing, promoting ethical AI development, and investing in reskilling programs to help the workforce adapt to an increasingly automated world. Preparing now will ensure societies can harness the potential of these technologies while mitigating their risks.

Pro Tip: One thing governments can do immediately is engage in mandatory 'digital safety training' (DST). Similar to workplace training for construction workers and factory employees, this new DST would cover everything from detecting scams and deepfakes to leveraging AI tools for maximum productivity and learning how to recognize digital malfeasance. In addition, this digital literacy could extend to how to protect your family from phishing attacks using personal 'safewords' that only you and your family know and never speak about near any digital device.

FUTURE OUTLOOK

Quantum Computing at 1M Qubits in 2030

By 2030, quantum computers are expected to reach the milestone of one million qubits, representing a significant leap from current quantum systems. At this scale, quantum computers will be capable of solving complex problems far beyond the reach of classical supercomputers, such as simulating molecular structures for drug discovery, optimizing vast logistical networks, and cracking previously unsolvable encryption. The development of quantum error correction will be crucial to reaching stable, high-performance systems, as qubits are prone to noise and errors. Quantum computing will revolutionize industries such as materials science, cryptography, and artificial intelligence, unlocking new levels of efficiency and innovation. Until we reach a stable quantum, other solutions, such as quantum simulators, will offer immediate value to users.

Artificial Intelligence - Point Solutions in 2030

By 2030, AI point solutions—narrow AI systems designed to solve specific tasks—will be deeply integrated into various industries. These solutions will automate specialized roles like detecting fraud, analyzing medical images, optimizing supply chains, and personalizing customer experiences. Point solutions will perform repetitive, data-driven tasks more efficiently than humans, freeing professionals to focus on more complex and creative responsibilities. These AI systems will continue to evolve, becoming faster and more accurate, and will be standard features in healthcare, finance, and manufacturing.

Artificial Intelligence - General Solutions in 2030

By 2030, general AI solutions, such as large language models (LLMs), will be far more powerful and capable of handling complex tasks across multiple domains. It is estimated that algorithms will be approximately one million times more potent than GPT-4, which was released in 2023 (Center for a New American Security). These AI systems will be able to understand, generate, and manipulate natural language with human-like proficiency, facilitating creative writing, marketing, sales, logistics, programming, coding, and data analysis. LLMs and other general AI tools will streamline operations across sectors like education, customer service, and content creation. They will be instrumental in bridging the gap between point solutions and more advanced forms of AI, enabling more sophisticated decision-making and problem-solving. Whether maximizing efficiency in a factory or ensuring the most effective customer service experience, AI will deliver way above expectations and way below cost estimates, making it a powerful new tool in every CEO's arsenal.

Artificial General Intelligence (AGI) in 2030

AGI refers to a level of AI that can perform any intellectual task better than any human can. By 2030, while AGI may still be in development, its emergence will be on the horizon, with AI systems approaching human-level understanding in certain areas. AGI can transfer knowledge from one domain to another, generalize learning across tasks, and exhibit reasoning abilities akin to humans. The development of AGI will raise profound ethical and societal questions, such as the role of AI in decision-making, the impact on jobs, and the potential risks of autonomous systems.

Generative AI for Creative, Middle Management, Computer Vision, Doctors, Lawyers, Accountants

By 2030, generative AI will play a significant role in fields requiring creativity and analytical thinking, such as content creation, middle management, and professional services like law and medicine. Generative AI will create original art, write reports, generate legal documents, and assist in medical diagnoses. Middle management tasks such as scheduling, project management, and performance analysis will be primarily automated by AI, allowing managers to focus on higher-level strategic decisions. AI will assist professionals like doctors, lawyers, and accountants by handling routine tasks, while humans will oversee complex, ethical, and judgment-based decisions.

WHAT HAPPENS WHEN AGI IS BETTER THAN MOST HUMANS AT MOST WORK?

When AGI surpasses human capabilities in most forms of work, it will transform the nature of labor and economic structures. Many jobs that require repetitive or analytical thinking will be fully automated, leading to potential job displacement. However, it will also create opportunities for new roles that involve supervising AGI, addressing ethical concerns, and overseeing AI's integration into society. Governments and industries will need to adapt quickly, focusing on reskilling workers, ensuring equitable wealth distribution, and redefining human labour's role in a highly automated world.

To adapt to these rapid changes, governments and industries will face the challenge of reskilling and upskilling a workforce in transition. This will involve large-scale educational efforts to provide displaced workers with the tools they need to navigate an AGI-driven economy, such as skills in human-AI collaboration, critical thinking, emotional intelligence, and creativity—qualities that are challenging for AGI to replicate. The emphasis on reskilling will ensure that individuals can take on roles that require uniquely human qualities and insights, preserving the role of human labour even in an automated world.

Artificial Super Intelligence - When Algorithms Become Smarter Than All Humans

Artificial Super Intelligence (ASI), where AI surpasses human intelligence across all domains, will be the most significant development in human history if achieved. ASI will be able to outthink, outlearn, and outstrategize the best human minds, posing existential questions about how society controls or collaborates with such a system. The development of ASI will require stringent safeguards, ethical frameworks, and oversight to ensure it acts in humanity's best interests. Humanity will need to adapt by rethinking governance, ethical decision-making, and global cooperation to ensure ASI benefits everyone and mitigates risks of misuse. For the first time in the history of the world, humans will no longer be the most intelligent species on the planet.

Future of Work in a Post-AGI World

In a post-AGI world, the nature of work will shift dramatically. Many routine and even skilled jobs will be automated, leading to widespread changes in employment structures. Human creativity, emotional intelligence, and complex problem-solving will become more valuable, as these are areas where AGI may only partially replace human capabilities. New industries will likely emerge around human-AI collaboration, where humans guide, supervise, or enhance AI's capabilities.

Economically, wealth distribution models must adapt to ensure that the productivity gains from AGI benefit society as a whole rather than creating further income inequality. Models like universal basic income (UBI) or wealth redistribution programs may become essential as traditional wages from full-time employment decline. This redistribution will support individuals whose work may not be needed in the same capacity, allowing them to thrive in a society that no longer relies on human labor for economic growth. With basic financial needs met, people may find fulfillment through creative, community-oriented, or intellectual pursuits rather than traditional jobs. We can pay people to spend time with retirees, capturing their stories and transferring real human wisdom to the next generations.

Ultimately, as AGI can perform many jobs, society's understanding of work and purpose may evolve. Human labor will likely shift toward roles prioritizing creativity, social connection, and innovation—areas where AGI, no matter how advanced, may struggle to replace the value of human experience and perspective fully. In a new, AGI-driven world, this transformation will push society to redefine work not as a necessity for survival but as a means for personal and collective fulfillment.

QUANTUM COMMUNICATIONS (ENTANGLED ONES AND ZEROS): INSTANT COMMUNICATION ANYWHERE IN THE UNIVERSE

By 2030, quantum communication technology, using quantum entanglement, will push the boundaries of secure, instantaneous communication. While accurate faster-than-light communication remains theoretical, quantum entanglement will allow for ultra-secure data transmission, particularly in financial transactions and government communications applications. Theoretically, entangled particles could enable communication across vast distances, including space, without intercepting traditional signals, revolutionizing secure communications for interplanetary exploration.

What Happens When Quantum Computers Can Crack Our Existing Security Protocols?

When quantum computers can crack current encryption protocols, which rely on classical computational difficulty, it will render much of today's digital security infrastructure obsolete. Governments, businesses, and individuals must use quantum-resistant encryption algorithms to protect sensitive data. The threat to existing financial systems, communications, and government infrastructure will necessitate global cooperation to develop and implement new cryptographic standards capable of withstanding quantum attacks.

One of the most promising applications of quantum entanglement is Quantum Key Distribution (QKD). QKD uses the principles of quantum mechanics to generate and distribute cryptographic keys. The key advantage of QKD is its ability to detect any eavesdropping attempts, as measuring a quantum state inevitably alters it, alerting the communicating parties to potential security breaches.

What Challenges Can Quantum Solve for Humanity?

Quantum computing will address some of humanity's most complex challenges by 2030. In healthcare, it will accelerate drug discovery and gene therapy by simulating molecular interactions at an atomic level. In climate science, quantum computing will model climate systems in greater detail, helping to develop practical solutions for mitigating climate change. It will also revolutionize logistics and optimization, solving previously unsolvable problems in global supply chains and energy distribution, ensuring more efficient resource use.

Accelerated drug discovery will allow researchers to simulate complex molecular interactions more accurately, leading to faster identification of potential drug candidates. Quantum-classical hybrid systems will enhance the prediction of chemical compound properties, removing approximations associated with classical computing methods. Integrating quantum computing with artificial intelligence will allow for testing therapeutic targets and drug modalities in larger quantities and at an accelerated pace. Quantum computing could also enhance our understanding of RNA structures and improve the accuracy of genetic sequencing and analysis, which is crucial for identifying potential drug targets and predicting interactions of small molecule drugs with RNA molecules.

Quantum computing will also advance climate science by enabling more complex and detailed climate system modelling, providing deeper insights into climate change mechanisms. This enhanced modelling capability will help scientists develop more effective solutions for mitigating climate change impacts. By 2030, quantum computing is expected to transform logistics and optimization across various industries by tackling previously unsolvable problems in supply chain management, potentially reducing waste and improving efficiency on a global scale. Complex routing and scheduling problems could be solved more effectively, significantly reducing transportation costs and emissions.

Quantum computing could also revolutionize energy grid management by optimizing distribution and reducing energy waste, enabling more efficient integration of renewable energy sources into existing power grids and supporting the transition to sustainable energy systems.

OPEN SOURCE AI VS. CLOSED SOURCE AI

The debate between open-source and closed-source AI is more critical than ever. Open-source AI promotes transparency, collaboration, and innovation, enabling developers worldwide to build and improve upon existing models. Closed-source AI, on the other hand, offers more control, security, and profitability for private companies but risks monopolizing powerful technologies. The balance between open and closed systems will determine how broadly AI's benefits are distributed and how much power is concentrated in the hands of a few corporations or governments. Society will need to weigh the benefits of innovation with concerns about security, fairness, and equitable access to AI technologies. This will be one of society's most significant challenges between now and 2030.

AI And The Abundance Economy

As the world grapples with the challenges of scarcity and limited resources, the concept of abundance economy has emerged as a beacon of hope for a more prosperous and sustainable future for individuals, businesses, and societies.

The abundance economy represents a fundamental shift away from the traditional scarcity mindset, where resources are perceived as limited, and competition drives economic growth. In contrast, the abundance economy is based on the idea that resources can be harnessed and regenerated to benefit everyone, fostering collaboration and technological innovation as the primary forces of prosperity. This paradigm is grounded in recognizing that many resources—energy, food, and even space exploration potential—are not inherently finite. With breakthroughs in renewable energy sources like nuclear fusion and advanced solar technology, humanity could achieve virtually unlimited clean power, enabling us to manage Earth's climate effectively. This abundance of energy would give us massive AI computing power and allow us to desalinate seawater, increase agricultural production, restore ecosystems, and support sustainable living, ultimately paving the way for ventures such as asteroid mining and deep space exploration. The abundance economy envisions a future where technology and innovation work hand-in-hand to create a world of shared prosperity and infinite potential.

EARLY DAYS OF AI (1950S-1970S)

The concept of AI dates back to the 1950s when computer scientists such as Alan Turing, Marvin Minsky, and John McCarthy began exploring the idea of creating machines that could think and learn like humans. The Dartmouth Summer Research

Project on Artificial Intelligence, held in 1956, is often credited as the birthplace of AI as a field of research.

The early years of AI were marked by significant breakthroughs, including the development of the first AI program, Logical Theorist, by Allen Newell and Herbert Simon in 1956. This program was designed to simulate human problem-solving abilities by using logical reasoning and search algorithms.

The AI Winter (1980s-1990s)

AI research faced significant challenges in the 1980s and 1990s despite the promising start. The field experienced declining funding and interest, often called the "AI winter." This period was marked by the failure of several AI projects, including the infamous ELIZA chatbot, which could not engage in meaningful conversations.

The Resurgence of AI (2000s-Present)

The early 21st century marked a powerful resurgence in artificial intelligence (AI), propelled by exponential advances in computing power, data storage, and breakthroughs in machine learning algorithms. In particular, developing deep learning techniques such as convolutional neural networks (CNNs) and recurrent neural networks (RNNs) revolutionized AI's capabilities by enabling machines to process vast datasets and learn complex patterns independently. CNNs became instrumental in image and video analysis, achieving remarkable performance in object detection, facial recognition, and autonomous driving. Meanwhile, RNNs enabled advancements in sequential data analysis, powering applications in natural language processing (NLP), speech recognition, and translation.

The rise of big data further fueled this resurgence, providing AI systems with the large, diverse datasets needed to train sophisticated models. This data influx allowed AI to outperform humans in specific tasks, such as image classification and strategic games, evidenced by the successes of AlphaGo and IBM Watson. Additionally, improvements in graphics processing units (GPUs) and the development of tensor processing units (TPUs) offered the computational power necessary to run deep neural networks on a massive scale, allowing AI models to process information and learn at speeds never seen before.

The resurgence was also marked by an evolution in unsupervised learning and reinforcement learning techniques, which allowed AI systems to make sense of unstructured data and learn complex decision-making strategies. In applications ranging from autonomous vehicles to healthcare diagnostics, AI has evolved from performing pre-programmed tasks to learning from experience, dynamically adapting to new situations, and optimizing solutions over time.

Current State of AI

Today, AI is an integral part of our daily lives, with applications in various industries, including:

- **Natural Language Processing (NLP):** AI-powered chatbots and virtual assistants, such as Siri, Alexa, and Google Assistant, have revolutionized how we interact with technology.

- **Computer Vision:** AI-powered image recognition and object detection are used in applications such as self-driving cars, autonomous mobility, precise agriculture, facial recognition, and medical diagnosis.

- **Robotics:** AI-powered robots are used in manufacturing, healthcare, and logistics, enabling automation and increasing efficiency. These robots will soon be doing various chores in your house.

- **Healthcare:** AI is used in medical diagnoses, treatment planning, and personalized medicine, improving patient outcomes and reducing costs. Systems can now analyze genetic profiles to tailor treatments, predict patient outcomes, and streamline medical processes, improving outcomes and reducing healthcare costs.

- **Finance:** AI-powered trading platforms and risk management systems analyze market trends and make data-driven investment decisions.

- **Entertainment and Media:** AI already powers personalized content recommendations on platforms like Netflix and Spotify by analyzing user preferences and predicting engagement.

- **Education:** AI plays a significant role in personalized learning, where systems can adapt to individual students' needs, offering customized content and pacing based on progress.

Challenges and Concerns

Despite the many benefits of AI, some several challenges and concerns need to be addressed:

- **Ethics:** AI systems can perpetuate and even amplify biases and discriminate against certain groups, highlighting the need for ethical considerations in AI development. Without robust ethical standards, AI may perpetuate existing inequalities or create new ones, disproportionately affecting marginalized communities. For instance, facial recognition systems have faced criticism for higher error rates among people of color, demonstrating the need for ethical frameworks and guidelines in AI development.

- **Job Displacement:** AI may displace specific jobs, particularly those involving repetitive tasks or tasks that can be automated. Roles in manufacturing, data entry, and customer service may significantly reduce, raising concerns about economic inequality and unemployment.

- **Data Quality:** The quality and accuracy of AI systems rely heavily on the quality and availability of data. Incomplete, biased, or inaccurate data can lead to erroneous or harmful outcomes, limiting the system's effectiveness. Ensuring high-quality, diverse, and representative datasets is crucial for building robust AI systems.

- **Explainability:** AI systems can be complex to explain and understand, making identifying and addressing biases and errors challenging. Explainability is particularly important in high-stakes domains such as healthcare, finance, and law, where understanding the rationale behind AI decisions is critical for trust and accountability. Research in interpretable AI and transparent algorithms is ongoing to make AI models more understandable and trustworthy.

- **Privacy and Surveillance:** AI systems often require extensive data collection to function effectively, raising concerns about privacy and the potential for intrusive surveillance. Advanced AI-powered monitoring tools can track an individual's behavior, location, and emotions, leading to a "surveillance society" where personal freedom is diminished.

- **Security and Cyber Threats:** AI introduces new cybersecurity risks, as adversaries may exploit vulnerabilities in AI systems or use AI to carry out sophisticated attacks. For example, deepfake technology can create realistic but false audio or video content, enabling identity theft, fraud, and misinformation.

- **Dependency and Skill Degradation:** As AI becomes increasingly capable of performing complex tasks autonomously, there is a risk that human skills and knowledge will erode over time. In fields where AI handles everything from routine tasks to high-level analysis, human expertise may diminish, leaving society vulnerable should AI systems fail or become unavailable. Don't let the algorithms make us dumber!

THE RISE OF THE ENTREPRENEURIAL SPIRIT

The digital and gig economy has given rise to a new generation of innovative entrepreneurs redefining the concept of work, allowing them to work anywhere in the world. This 'digital nomad' class will expand as businesses do mass layoffs caused by AI outperforming than humans in some jobs. AI enables individuals to monetize their skills and services and create their businesses. This has already led to a rise in entrepreneurship and innovation, as individuals are empowered to pursue their passions and create their own opportunities. This will expand as we get closer to 2030.

The gig economy, characterized by short-term, flexible, and often freelance work arrangements, has become a dominant feature of the modern labor market and allows people to:

- **Increase Flexibility:** The gig economy has given rise to a new level of flexibility and autonomy in the workforce, as individuals can choose their work arrangements and schedules.

- **Learn New Skills and Competencies**: The gig economy has given rise to new skills and competencies, such as data analysis and programming, which are in high demand, although AI is starting to code better than humans.

- **Create New Business Models:** The gig economy has created new business models, such as ride-sharing and food delivery, disrupting traditional industries.

The gig economy has also increased income inequality, as those with the skills and resources to adapt to the new economy are more likely to thrive. It really will be Humans + AI = Awesome!

Communication

On a crisp autumn morning in the 2030s, Naomi, a digital health specialist, is on her way to her office in San Francisco. At the same time, her partner Rizwan, a software engineer based in Kenya, started his day remotely from a sunlit café in Nairobi. Living 10,000 miles apart, they still felt connected thanks to the revolutionary advancements in communication technology.

Their day began with a 6G-enabled video call, where latency is nearly nonexistent, allowing them to communicate as if they were in the same room. Through non-invasive brain-machine interfaces (BMIs), they can quickly exchange ideas, transferring not just words but data files and images directly from their devices to each other's screens. This innovation has transformed Naomi's work in healthcare, allowing her to instantly share detailed 3D patient data and treatment plans with specialists across continents. For Rizwan, these advanced communication tools mean that collaboration with his global team flows smoothly, with files and commands transmitted almost as quickly as they are created.

Later, Naomi headed into her office, where she used XR-powered virtual meetings to consult with patients worldwide, who accessed her guidance from their homes without worrying about travel. She reflected on how this communication technology has bridged vast distances, especially in underserved rural areas where access to healthcare has traditionally been limited. Thanks to ubiquitous satellite internet coverage, even in remote regions, Naomi's services can reach those who once had limited or no access to specialized medical care.

Back in Nairobi, Rizwan attended a virtual conference on the Metaverse, a platform that has evolved beyond social interactions into a virtual space for professional work and economic exchange. As he navigated his avatar through the Metaverse, Rizwan interacted with developers from Japan to Brazil, exchanging ideas and resources with the fluidity of a face-to-face meeting. The Metaverse has made international collaboration more engaging and immersive than ever before, allowing people to connect and innovate without the constraints of geography.

In the afternoon, Naomi received a notification from her AI assistant, which predicted that a patient might need attention based on real-time health metrics gathered through 5G and 6G IoT devices. Meanwhile, Rizwan's AI-driven programming assistant flagged potential bugs in his code, streamlining his workflow and minimizing errors. Their respective fields, healthcare and software engineering, are deeply interconnected in ways they could not have imagined a decade ago, thanks to the seamless, hyper-connected communication infrastructure.

After their workday ended, Naomi and Rizwan used telepresence technology to enjoy a virtual dinner together, discussing their days over an immersive XR experience that recreated the ambiance of their favourite restaurant. These advancements in predictive and personalized communication technologies have simplified their professional lives and allowed them to maintain a meaningful connection despite the distance.

For Naomi and Rizwan, 2030 isn't just about faster communication—it's about breaking down the last of the digital and geographical barriers, empowering them to work, create, and connect with the world in unprecedented ways.

COMMUNICATION IN 2030

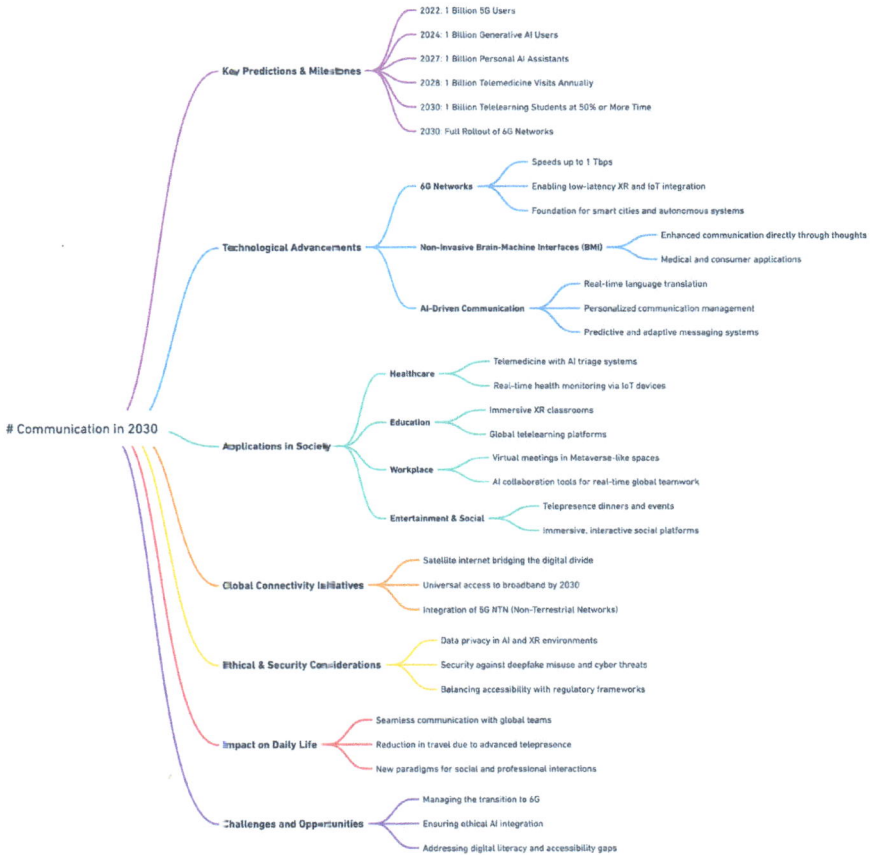

- # Communication in 2030
  - Key Predictions & Milestones
    - 2022: 1 Billion 5G Users
    - 2024: 1 Billion Generative AI Users
    - 2027: 1 Billion Personal AI Assistants
    - 2028: 1 Billion Telemedicine Visits Annually
    - 2030: 1 Billion Telelearning Students at 50% or More Time
    - 2030: Full Rollout of 6G Networks
  - Technological Advancements
    - 6G Networks
      - Speeds up to 1 Tbps
      - Enabling low-latency XR and IoT integration
      - Foundation for smart cities and autonomous systems
    - Non-Invasive Brain-Machine Interfaces (BMI)
      - Enhanced communication directly through thoughts
      - Medical and consumer applications
    - AI-Driven Communication
      - Real-time language translation
      - Personalized communication management
      - Predictive and adaptive messaging systems
  - Applications in Society
    - Healthcare
      - Telemedicine with AI triage systems
      - Real-time health monitoring via IoT devices
    - Education
      - Immersive XR classrooms
      - Global telelearning platforms
    - Workplace
      - Virtual meetings in Metaverse-like spaces
      - AI collaboration tools for real-time global teamwork
    - Entertainment & Social
      - Telepresence dinners and events
      - Immersive, interactive social platforms
  - Global Connectivity Initiatives
    - Satellite internet bridging the digital divide
    - Universal access to broadband by 2030
    - Integration of 5G NTN (Non-Terrestrial Networks)
  - Ethical & Security Considerations
    - Data privacy in AI and XR environments
    - Security against deepfake misuse and cyber threats
    - Balancing accessibility with regulatory frameworks
  - Impact on Daily Life
    - Seamless communication with global teams
    - Reduction in travel due to advanced telepresence
    - New paradigms for social and professional interactions
  - Challenges and Opportunities
    - Managing the transition to 6G
    - Ensuring ethical AI integration
    - Addressing digital literacy and accessibility gaps

COMMUNICATION PREDICTIONS

- 2022: One billion 5G users by the end of year

- 2024: One billion people using digital fitness & well-being devices and services

- 2024: One billion people using generative AI

- 2026: One billion on-device AI users

- 2027: One billion personal AI assistants (including medical) and AI agents

- 2027: One billion VR users using an immersive device one hr/day

- 2028: One billion telepresence humans; 25% of activities being done virtually that were previously done with travel

- 2028: One billion telemedicine medicine visits per year

- 2028: One billion students having access to >25 mbps broadband service

- 2029: One billion humans with non-invasive brain-machine interface (BMI)

- 2030: One billion students telelearning at least 50% of the time

- 2033: Five billion personal AI assistants (including medical case management) and AI Agents

- 2033: One billion 6G devices

- 2039: One billion humans with invasive brain-machine interface (BMI)

- 2043: One billion 7G devices

- 2050: Everyone will have a communication device because communication and information will be too cheap to meter

WHAT IS GOING TO HAPPEN?

By 2030, communication technologies will transform everyday life through ultra-fast and ubiquitous connectivity. The rollout of 6G networks will further enhance internet speeds, reduce latency, and expand network reliability, profoundly impacting how we interact, work, and consume information. Advances in AI will personalize communication tools, making them more intuitive and integrated in our daily routines. Additionally, emerging technologies like XR and non-invasive brain-machine interfaces will begin to blur the lines between digital and physical worlds, offering new ways to interact with our environment and each other, thereby reshaping social interactions and creating more immersive, efficient, and tailored experiences. We will no longer look down at our smartphones but look out into the world with the aid and filters of XR and non-invasive brain-machine interfaces. It will change how we look at the world in profound ways.

WHY IT MATTERS?

The evolution of communication technologies is pivotal as it directly influences economic growth, social integration, and global connectivity. Enhanced communication systems like 5G and upcoming 6G networks matter because they enable rapid, real-time information exchange, vital for innovation and crisis response. These advancements democratize access to technology, reducing the digital divide and connecting underserved and rural areas. Furthermore, they are foundational to developing smart cities and implementing IoT solutions, leading to more sustainable and efficient urban environments. Ultimately, communication technologies drive the modern economy and shape the dynamics of human interactions, making their progression essential for future societal development.

WHO WILL IT IMPACT?

The sweeping advancements in communication technology will impact virtually everyone. Notably, it will transform the lives of those in rural and underserved regions by providing reliable, high-speed internet access, thus bridging the digital divide. It will also significantly affect industries reliant on real-time data, such as healthcare, finance, and transportation, enhancing service delivery and operational efficiencies. Educators and students will benefit from improved access to digital learning resources, fostering global education equity. Moreover, businesses across all sectors must adapt to these technologies to remain competitive, making the impact of communication innovations both widespread and profound.

CALL TO ACTION

To prepare for the rapid advancements in communication technology by 2030, the most crucial action individuals and organizations can take is to invest in digital literacy and infrastructure. Embracing continuous education and training in emerging technologies will enable adaptability to new communication tools and platforms as they evolve. For businesses, this means upgrading systems and fostering a culture that embraces technological change. Staying informed about trends and potential regulatory changes in telecommunications will ensure that all stakeholders can leverage these technologies effectively and ethically, maintaining relevance in an increasingly connected world. Ensure you are focused on cybersecurity so that the communication does not do more harm than good.

DOUG HOHULIN ON COMMUNICATIONS

In 1989, my cellular communication journey with Motorola began as a 1G cellular system engineer working in Japan, marking the start of an adventure through the evolving landscapes of cellular technology. In 2011, Nokia-Siemens purchased

Motorola Network, which is how I came to Nokia. From the nascent stages of 1G to the development of 6G technology, my career until 2021 was pursuing what comes next in digital communication.

I transitioned from engineering to business development and strategy roles and started presenting about the future of technology. My first public presentation on 3G technology was in the early 2000s in Hawaii. I would continue delivering on the future of technology topics of Bridging the Digital Divide, 4G, 5G, 6G, XR/Metaverse, Digital Health, AI, Autonomous Vehicles, and Cybersecurity and gave public presentations to industry groups (IWPC, IEEE, Western 95 Show, SCTE meetings and Cable96) and Universities, High schools and even 4th graders. My career has been a continuous quest to uncover the "next generation" of technology – especially communication-related. However, my passion extends beyond mere technological advancements. I am deeply invested in the potential of technology to bridge gaps, connect the unconnected, and foster an era of abundance for an exponential leap for humanity.

The 5Gs of Cellular Communication

There have been 5Gs of cellular communication deployed to date:

- 1G – Analogue cellular

- 2G – Digital cellular

- 3G – Mobile broadband, deployed in the 2000's

- 4G – LTE: deployed in the 2010s

- 5G – Deployed in the 2020s that provides:

    - Extreme mobile broadband with peak speeds of >10 Gbps and 100 Mbps where needed

- Massive machine communication for low-cost IoT to support 10s of billions of devices

- Critical machine communication is used for applications that require ultra-reliability, such as 5G transportation applications.

Highlighted on the Qualcomm site and HIS report highlighted, "5G will lift mobile into a technology that changes the world. ... In 2035, when 5G's full economic benefit should be realized across the globe, a broad range of industries – from retail to education, transportation to entertainment, and everything in between – could produce up to $12.3 trillion worth of goods and services enabled by 5G mobile technology" (https://www.qualcomm.com/invention/5g/economy)

Reflecting on the monumental advancements since the early days of 1G in the 1980s, consider how each generation of communication technology has progressively amplified humanity's ability to connect, share, and innovate. From the humble beginnings of analog signals to the high-speed capabilities of 5G and beyond, we've seen a transformation in how we communicate and live our daily lives.

A Brief History of Mobile Communications

As we delve into the milestones of mobile communication technologies, Consider the cultural, economic and technological shifts that facilitated this rapid adoption and the integration of these technologies into the fabric of modern society over the last two decades.

The Advent of One Billion Users of Technology:

- 2002 – One billion cell phone users

- 2005 – One billion internet users

- 2008 – One billion installed PC based

In 2002, the world celebrated the achievement of one billion cell phone users—precisely 200 years after humans reached the 1 Billion milestone, a testament to the rapid adoption of technology across diverse demographics. The first 3G network was launched in Japan on October 1, 2001, and people could use their cell phones to connect to the internet. This milestone underscored the mobile phone's evolution from a luxury item to an indispensable tool for communication, offering unprecedented access to information and services.

Then, in 2005, the internet reached its billion-user milestone, further democratizing access to information and connecting individuals worldwide. The synergy between mobile technology and the internet began to erase geographical and socio-economic barriers.

One billion PC users were reached in 2008, marking another critical point in the digital era, emphasizing the role of personal computing. Microsoft's original mission was to put "a computer on every desk and in every home," which became a reality. Now, the world could envision a computer or mobile phone for every person.

These milestones in the advent of the internet and mobile communication heralded a new age of how people will engage with technology.

2011: The Milestone of One Billion Smartphone Users

In 2011, the world witnessed a landmark technological moment: the number of smartphone users eclipsed the one billion mark. This milestone was not merely a quantitative measure but symbolized a significant transformation in global communication and digital connectivity. It marked the advent of an era where smartphones became central to daily life, offering unprecedented access to information, entertainment, and services on the go.

This pivotal year also represented a transitional period in mobile network technology as the world shifted from 3G to 4G. The leap to 4G technology was crucial, as it offered faster internet speeds, reduced latency, and enhanced capacity, allowing for a richer, more seamless mobile internet experience. This transition was instrumental in catering to the growing demand for mobile data driven by the surge in smartphone adoption.

The introduction of 4G technology amplified the capabilities of smartphones, making it possible to stream high-definition videos, engage in video calls, and access cloud services effortlessly. This shift not only elevated the user experience but also spurred innovation across various sectors, including mobile commerce, online education, and remote work, underscoring the role of smartphones as indispensable tools for modern living.

Moreover, the democratization of smartphone technology, facilitated by the rapid advancements and the availability of affordable models, played a critical role in reaching the one billion user milestone. It highlighted the smartphone's evolution from a luxury item to a necessity, bridging the digital divide and offering billions worldwide a gateway to the digital economy.

2014: Number of Mobile Devices = Human Population

In 2014, a transformative milestone was reached in digital technology: the number of mobile devices in circulation globally equaled the human population. This momentous occasion signified the ubiquity of mobile technology and underscored its integral role in shaping human interaction, communication, and access to information. This equilibrium point between devices and humans marked a pivotal shift towards a profoundly interconnected global society, where mobile devices became extensions of the human experience.

The significance of this milestone goes beyond the sheer volume of devices; it reflects the rapid advancement and adoption of mobile technology across diverse socio-economic backgrounds. Mobile devices, encompassing smartphones and tablets, became the primary means of internet access for many, democratizing information and services like never before. They bridged geographical and cultural divides, providing a platform for social inclusion, economic opportunity, and educational resources.

Furthermore, the parity between mobile devices and the human population in 2014 highlighted the potential for mobile technology to drive innovation in healthcare, education, and finance, particularly in remote or underserved regions. This milestone underscored the transformative power of mobile technology, not just as a tool for communication but as a catalyst for societal change and development.

The Significance of Five Billion Smartphone Users and Internet Users in 2022 and One Billion 5G Users:

The launch of the first 5G network at the end of 2018 marked the beginning of a new era in telecommunications, showcasing the rapid pace at which technology can evolve and be adopted globally. By 2022, just four years after its introduction, the world saw 1 billion 5G users, a testament to the exponential growth potential of new technologies once the foundational infrastructure, like 4G, is in place. This swift expansion reflects the demand for faster and more reliable internet connectivity and the readiness of consumers and industries to embrace and integrate advanced technological capabilities into daily life.

2022 – Five Billion Humans Using a Smartphone

In 2022, the global count of smartphone and internet users reached an astounding milestone of five billion users, underscoring the monumental role these devices play in the fabric of daily life worldwide to the masses. This figure represents a significant portion of the global population, firmly establishing smartphones as the most widespread form of digital technology. The reach of smartphones, transcending geographical, cultural, and socioeconomic barriers, highlights a pivotal moment in the digital era, emphasizing the device's integral role in connecting humanity.

Going from one billion to five billion humans using these technologies demonstrates that this technology is not just for the developed world but the entire world. The significance of five billion smartphone users is multifaceted, touching on various aspects of modern society. Economically, it has spurred growth across multiple sectors, from mobile commerce and digital advertising to app development and telecommunications, creating millions of jobs worldwide. The proliferation of smartphones has also catalyzed innovation, leading to the development of apps and services that cater to many needs and preferences, from health and education to entertainment and finance.

Socially, smartphones have redefined how individuals communicate, interact, and form communities. They have enabled real-time communication and access to information on an unprecedented scale, fostering a global exchange of ideas and cultures. Moreover, smartphones have become critical tools in mobilizing social and political movements, allowing for the organization and dissemination of information like never before.

From a technological standpoint, reaching five billion users has driven mobile technology advancements, including hardware improvements, such as more powerful processors, enhanced camera technology, and software with more intelligent, intuitive operating systems and applications. This milestone has also pushed forward the development of mobile internet infrastructure, including expanding 4G and 5G networks, to support the increased demand for data and connectivity.

In essence, the milestone of five billion smartphone users in 2022 is a testament to the device's ubiquity and a reflection of its profound impact on society. It highlights the smartphone's role as a key driver of the ongoing digital transformation, influencing every aspect of modern life and shaping the future of global connectivity.

The Role of 5G in Rural and Underserved Areas

The deployment of 5G technology in rural and underserved areas heralds a transformative era in global connectivity, promising to bridge the stark digital divide that has long marginalized these communities from the mainstream digital economy. The introduction of 5G networks is set to revolutionize these areas by providing high-speed, reliable internet access, a critical foundation for fostering educational opportunities, healthcare accessibility, and economic development.

Rural and underserved regions have grappled with limited or non-existent internet connectivity for decades, significantly hindering their development and access to essential services. Traditional broadband solutions often need to catch up due to the high costs and logistical challenges of extending infrastructure to remote areas. However, 5G technology, with its ability to deliver faster internet speeds over broader areas with lower latency, presents an unprecedented opportunity to overcome these barriers.

The impact of 5G in these regions extends far beyond just enhanced internet access. In education, it enables remote learning platforms and digital classrooms, offering students in remote areas access to quality educational resources and global knowledge bases. In healthcare, 5G facilitates telemedicine services, allowing patients in rural areas to consult with specialists and receive medical advice without the need to travel long distances. This is particularly vital for chronic disease management and emergency medical consultations, where timely intervention can be life-saving.

Moreover, 5G technology plays a pivotal role in economic empowerment by enabling rural businesses to tap into digital marketplaces, access online banking services, and leverage digital tools for agriculture, such as precision farming, which significantly increases productivity and sustainability. It also opens up new avenues for innovation and entrepreneurship in these communities, driving local development and integration into the global economy.

However, successfully implementing 5G in rural and underserved areas requires concerted efforts from governments, telecom operators, and international organizations. Policies and investments are needed to support infrastructure development, reduce deployment costs, and ensure that the benefits of 5G technology are accessible to all segments of society. Furthermore, initiatives to enhance digital literacy and skills among rural populations are crucial to maximizing the potential of 5G technology.

The advent of 5G technology from space is a groundbreaking development for global connectivity, particularly for rural and underserved areas. This innovation aims to deliver high-speed internet access worldwide, including regions traditionally challenged by connectivity issues, by leveraging non-terrestrial networks (NTN). Such advancements could significantly enhance everyone's access to information, education, healthcare, and economic opportunities, marking a significant leap toward digital inclusivity. Qualcomm's discussion on this topic "5G from space: The final frontier for global connectivity" provides further insights.

The deployment of 5G in rural and underserved areas stands at the cusp of driving significant social and economic change, offering a lifeline to global inclusivity and connectivity. By addressing the challenges and leveraging the opportunities presented by 5G, we can look forward to a future where geographical location no longer dictates access to the digital world, ensuring that every community is included in the ongoing digital revolution. The journey towards this future is complex and fraught with challenges. Still, the potential rewards for rural and underserved communities worldwide are immense, making it a pivotal focus for global development strategies.

COMMUNICATION FOR SPATIAL COMPUTING/XR/ METAVERSE & AI INNOVATIONS IN THE NEXT 5 YEARS

As we move towards 2030, the communication landscape is set to be revolutionized by integrating spatial computing, extended reality (XR), and the Metaverse, coupled with advancements in artificial intelligence (AI). These innovations are poised to transform how we interact with digital content and each other, offering new dimensions of connectivity and immersive experiences. Over the next five years, we can expect significant developments in these areas, reshaping the boundaries of digital interaction.

Spatial Computing: Enhancing Interaction with the Physical World

Spatial computing is at the forefront of merging the physical and digital worlds. This technology enables devices to perceive and interact with their environment in three dimensions. Over the next five years, spatial computing will become more integrated into daily tasks and operations across various manufacturing, healthcare, and real estate industries. For example, in manufacturing, spatial computing can assist in complex assembly processes by providing workers with real-time, augmented overlays of assembly instructions, reducing errors and increasing efficiency.

XR: Broadening the Horizon of Immersive Experiences

Extended Reality (XR) encompasses virtual reality (VR), augmented reality (AR), and mixed reality (MR), offering immersive experiences that blend real and virtual worlds. In the next five years, XR technologies are expected to advance significantly, driven by improvements in hardware, such as lighter, more comfortable headsets with higher resolutions and faster processing capabilities. These enhancements will broaden the use of XR in fields like education, where students can explore virtual labs or historical sites, and telecommuting, where virtual workspaces become more interactive and engaging, replicating the dynamics of physical office environments.

The Metaverse: A New Platform for Digital Economy and Social Interaction

The Metaverse, a collective virtual shared space created by converging virtually enhanced physical and digital reality, is anticipated to expand dramatically. It promises to offer a platform where users can work, play, and connect in ways that go beyond what current social media and online platforms allow. Over the next five years, we'll see the Metaverse becoming a hub for digital economies, where users can create, buy, and sell goods and services within a fully functioning economy powered by blockchain and cryptocurrency technologies.

AI Innovations: Driving Smarter, More Adaptive Environments

AI's role in supporting spatial computing, XR, and the Metaverse cannot be understated. AI will help these technologies understand and predict user behaviours, personalize experiences, and make interactions more intuitive. For instance, AI can enable adaptive learning environments in XR, adjusting educational content to suit individual learning speeds and styles. In the Metaverse, AI-driven avatars and agents could provide user support, manage transactions, and ensure security.

Over the next five years, as these technologies converge and mature, they will redefine the essence of how we perceive and interact with the digital world. The fusion of spatial computing, XR, and the Metaverse, enhanced by AI, will unlock unprecedented opportunities for creativity, efficiency, and connectivity, setting the stage for the next giant leap in communication technology.

PREDICTIONS

2027: One Billion Spatial Computer Metaverse Users:

In 2020, I presented to EducatorsInVR.com, "When will we reach 1 Billion 5G Users? 1 Billion VR/AR Users? Learning from the History of the 1st Billion Cell Phone & Smartphone" and in 2022, I gave a similar presentation to StudentsInVR based on a LinkedIn article I wrote at the end of 2019, "The 21st Century – The Century of the Billion & Billions."

I predicted that one billion people will be using these immersive devices at least one hour a day by 2027, within two years (my current estimate is 2028-2029). Then, five years later, this will be five hours a day and five billion people. Smart glasses and other immersive devices will replace the smartphone.

Communication Predictions for 2030 and Beyond

By 2030, the communication landscape is poised for transformative shifts driven by relentless technological advancements. These predictions for 2030 and beyond highlight the evolution of existing technologies and the emergence of new paradigms that will reshape how we connect, collaborate, and interact in a digital-first world.

The Maturation of 6G Technology

By 2030, 6G technology is expected to have begun its rollout, offering even faster speeds, lower latency, and greater capacities than 5G. 6G will enable incredibly high data rates, potentially up to terabits per second, revolutionizing how data is transmitted across networks. This will enhance immersive technologies such as virtual and augmented reality, making them part of everyday life. Additionally, 6G will facilitate the proliferation of smart cities and autonomous vehicles, where instantaneous communication (limited by the speed of light) and vast data transfer are crucial.

Ubiquitous Internet Connectivity

The dream of global internet coverage is likely to be realized by 2030, driven by satellite internet services and expanded 5G and 6G infrastructures. Companies like SpaceX are poised to deliver high-speed internet access to the most remote corners of the world, democratizing access to information and bridging the digital divide in unprecedented ways.

The Starlink Business: Direct To Cell website highlights how NTN will connect the unconnected; on January 2, 2024, SpaceX launched six Starlink satellites with Direct to Cell capabilities, completing initial tests successfully. On January 8, the first text messages were sent and received using unmodified cell phones, proving the system's functionality. This innovation enhances Starlink's mission by offering global connectivity for text, voice, and data on LTE devices. Text service begins in 2024, with voice, data, and IoT services planned for 2025. Building a satellite network that connects directly to standard cell phones poses unique technical and regulatory challenges beyond deploying Starlink's extensive constellation of over 5,000 satellites serving four million users (Sept. 2024) worldwide. Future launches via Starship will enhance the service and increase launch capacity.

Advancements in Brain-Computer Interfaces (BCI) (Non-invasive and Invasive)

Non-invasive BCI harnesses the brain's electrical signals without requiring surgical implants, making them more accessible and less risky. These devices use sensors placed on the scalp to detect brainwave patterns and translate them into commands that can control computers, prosthetics, or other connected devices. Advances in this technology are rapidly enhancing its accuracy and speed, allowing users to perform tasks ranging from moving virtual objects to operating wheelchairs just by thinking. As BCIs become more refined, they offer profound possibilities for individuals with mobility or communication impairments, promising greater independence and enhanced quality of life.

Invasive brain-computer interfaces connect the human brain to external devices and are expected to advance significantly by 2030. These devices will facilitate not only medical therapies and enhancements, such as restoring function to individuals with disabilities but also the broader consumer adoption for controlling devices through thought and accessing information directly through brainwaves. This could lead to a new human-computer interaction paradigm, making keyboards and touchscreens obsolete.

Integration of AI in Personal and Professional Communication

Artificial intelligence will become deeply integrated into both personal and professional communication spheres. AI will tailor communication tools to individual preferences and contexts, manage inboxes, and even simulate human conversation for tasks ranging from customer service to personal companionship. Advanced AI could also translate languages in real time, breaking down global communication barriers more effectively than ever.

Evolution of the Metaverse

The concept of the Metaverse will evolve beyond gaming and social interactions to become a platform for professional work, education, and commerce. Virtual environments will host international conferences, global classrooms, and virtual marketplaces. The Metaverse will provide a seamless, integrated virtual experience replicating and extending real-world activities.

Personalized and Predictive Communication Technologies

Communication technologies will be reactive and predictive, using AI to anticipate needs and provide information before it is explicitly requested. This could manifest in predictive texting evolving into predictive communication, where your devices anticipate and execute communications based on learned preferences and patterns.

By 2030, these advancements will redefine the essence of communication. They will enhance how we interact with each other and fundamentally alter the interfaces between humans and machines, paving the way for a future where connectivity is as natural and essential as the air we breathe. As we move towards this future, it is crucial to navigate these advancements responsibly, ensuring they are used ethically and to benefit all humanity.

Semiconductors and Data Centers

Werner leaned against the railing of his Arctic data center's observation deck, staring at the endless pristine snow and ice expanse. The hum of the facility behind him was a constant reminder of the monumental operation he helmed. This nuclear-powered data center drove the computational backbone for some of the most advanced AI systems in the world. The air was crisp, biting even, but Werner found it invigorating. Today wasn't just another day; it marked a critical breakthrough in a collaboration he was particularly proud of.

Half a world away, Maggie was engrossed in her work at her hyperscaler company's innovation lab in Singapore. The lab buzzed with quiet intensity, its walls lined with holographic displays showing complex data streams. As the head of data operations, Maggie was at the forefront of leveraging artificial general intelligence (AGI) to decode some of humanity's most complex problems. Her focus today? A real-time simulation of protein folding that could potentially unlock a cure for neurodegenerative diseases like Alzheimer's.

The connection between Maggie's groundbreaking work and Werner's infrastructure was seamless but pivotal. Her AGI models required staggering amounts of computational power to run the simulations. With its unrivalled efficiency and quantum-enhanced cooling systems, the nuclear-powered data center in the Arctic provided the necessary muscle to process these models at unimaginable speeds.

As Maggie prepared for a midday virtual check-in with Werner, her glasses chimed softly to alert her. The holographic display of Werner's Arctic office materialized in front of her desk. "Maggie," Werner said, his tone warm but professional, "your team's simulation output is impressive. Our engineers monitored the load; we've never seen the processors handle this complexity so smoothly. How are things progressing?"

Maggie smiled. "We've just completed a protein interaction map that would have taken five years of manual computation in 2020. We did it with your center's capabilities in less than a day. It's not just about speed—it's the precision. We're narrowing down viable pathways for drug development at an unprecedented rate."

Werner nodded, pleased. "That's the beauty of where we're headed. The efficiency of our nuclear facility, combined with the integration of quantum processors, means we're not just meeting the demands of today—we're future-proofing innovation."

Outside their conversation, the world was already changing because of the technology they drove. Maggie's hyperscaler company had partnered with global health organizations to accelerate research into neurodegenerative diseases and rare genetic disorders. The AGI systems, trained on vast datasets, identified patterns no human could see, uncovering potential treatments and prevention strategies.

Meanwhile, Werner's data center was more than just a technological marvel. It embodied a new era of sustainability. By harnessing nuclear power, his operation had significantly reduced its carbon footprint compared to traditional data centers. The waste heat generated by the facility was repurposed to power nearby Arctic greenhouses, growing fresh produce in one of the most inhospitable regions on Earth. It was a testament to how technological advancement and environmental responsibility could coexist.

Later that day, Maggie hosted a meeting in her virtual workspace, a photore-alistic replica of a lush tropical forest she often retreated to for inspiration. Her team of international researchers logged in as avatars, collaborating across time zones without the need for travel. "Thanks to Werner's upgrades," she told her team, "we've allocated more resources to a new initiative—mapping the cellular effects of long-term microgravity on the human body. This could revolutionize space medicine."

As the sun dipped below the Arctic horizon, Werner reflected on the broader implications of his work. The technology powering Maggie's AGI wasn't just solving problems but redefining what was possible. From enabling real-time medical breakthroughs to supporting global food security, the ripple effects were vast.

That evening, Maggie and Werner paused to savour small moments of gratitude. Maggie, for the ability to dream big and see those dreams materialize through the tools at her disposal. Werner, for building the infrastructure that made those dreams possible. Together, they were part of a larger story of humanity pushing past the present's limits to create a brighter, cleaner, and more connected future.

By 2030, data centers are projected to use around 8-10% of the world's power, based on current growth trends. With the explosion of cloud computing, AI processing, and the growing demand for data storage due to technologies like IoT and 5G, the energy consumption of data centers is expected to increase significantly. Efforts to counter this include advancements in energy-efficient hardware, AI-driven power management, and using renewable energy sources to power data centers. However, the sheer demand for processing and storage will still push data centers to consume more global electricity.

SEMICONDUCTORS and DATA CENTERS IN 2030

 # Semiconductors & Data Centers in 2030

- **Predictions**
  - Semiconductors reach sub-1nm scale, with 0.5nm chips becoming a reality
  - Data centers to consume 8-10% of the world's power
  - GPUs 10,000 times more powerful, enabling petaflop/zettalop AI inference
  - 3D chip stacking becomes widespread
  - AI-driven chip design creates a virtuous cycle for innovation

- **What is it?**
  - Semiconductors
    - Foundational for digital devices (smartphones, laptops, AI servers, quantum computers)
    - Enabling exponential computing power
  - Data Centers
    - Manage vast data oceans
    - Backbone of cloud computing, AI, IoT ecosystems
    - Evolving with green energy & edge computing

- **Applications of Semiconductors**
  - Computing
  - Communication
  - Artificial Intelligence
  - IoT devices
  - Medical devices
  - Automotive systems
  - Aerospace systems

- **Advancements by 2030**
  - Sub-1nm semiconductors for higher energy efficiency
  - 3D chip stacking for faster data movement & improved heat management
  - Quantum materials like graphene and carbon nanotubes
  - GPUs achieving zettaflop-level processing power
  - Liquid cooling & immersion cooling mainstreamed

- **Why It Matters**
  - Drives AI, IoT, and 5G applications
  - Backbone of digital infrastructure
  - Key to sustainable cloud computing
  - Supports quantum computing & scientific breakthroughs

- **Impacts**
  - Tech giants enhance cloud services
  - Hardware manufacturers innovate for higher efficiency
  - Consumers experience faster, smarter devices
  - Automotives advance autonomous technology
  - Healthcare achieves precision diagnostics & robotic surgeries
  - Telecommunications deploy 6G networks
  - Governments tackle energy sustainability for data centers

- **Call to Action**
  - Invest in clean energy solutions for data centers
  - Foster incentives for advanced semiconductor production
  - Adopt AI-driven cooling and optimization systems
  - Promote edge computing integration

PREDICTIONS

- Semiconductors will reach a sub-1nm scale, with 0.5nm chips becoming a reality.

- Data centers will consume 8-10% of the world's power.

- GPUs will become 10,000 times more potent than current models.

- 3D chip stacking will be widespread.

- AI-driven chip design will create a virtuous cycle where AI systems design better semiconductors, powering more advanced AI systems and accelerating technological progress across all fields.

WHAT IS IT?

Semiconductors will continue to be the lifeblood of our digital world, serving as the foundational technology behind everything from smartphones and laptops to the most advanced AI servers, quantum computers, and global data centers. These tiny yet powerful chips are the unsung heroes enabling the exponential growth of computing power and data processing capabilities.

At the same time, data centers will solidify their position as the nerve centers of our hyperconnected world, managing the vast oceans of data generated by billions of devices, sensors, and systems every second. These facilities will process, store, and transmit information at speeds and scales that were unimaginable just a decade ago, providing the computational backbone for technologies like autonomous vehicles, real-time language translation, and immersive metaverse experiences.

Driving this evolution are groundbreaking advancements in semiconductor technology. Thanks to innovations like sub-nanometer processes, 3D chip stacking, and new materials like graphene, chips will become even smaller, denser, and more efficient. These improvements won't just make devices faster; they'll also drastically reduce energy consumption, addressing the growing environmental concerns tied to data infrastructure.

APPLICATIONS OF SEMICONDUCTORS

- **Computing:** Semiconductors are the backbone of modern computing, powering everything from personal computers to smartphones and servers.

- **Communication:** Semiconductors are used in various communication devices, including routers, switches, and modems.

- **Artificial Intelligence:** Semiconductors are used in AI-powered devices like neural networks and machine learning algorithms.

- **Internet of Things (IoT):** Semiconductors are used in IoT devices, such as smart home devices, wearables, and sensors.

- **Medical Devices:** Semiconductors are used in medical devices, such as pacemakers, insulin pumps, and diagnostic equipment.

- **Automotive:** Semiconductors are used in automotive systems, including anti-lock braking, airbags, and navigation systems.

- **Aerospace:** Semiconductors are used in aerospace applications, including satellite communications and navigation systems, communication systems, and propulsion systems.

For data centers, these advancements mean scaling up capacity without exponentially increasing power demands. Imagine servers capable of handling petabytes of data in seconds, powered by high-performance chips optimized for AI and quantum computing. These data centers will incorporate cutting-edge technologies like liquid cooling, renewable energy sources, and edge computing to enhance efficiency, minimize latency, and meet the demands of a global digital economy.

By 2030, semiconductors and data centers will not just power devices and applications—they'll define the future of innovation, enabling breakthroughs across industries, from personalized medicine to climate modelling, while supporting the seamless connectivity and computational power that humanity relies on daily.

WHAT IS GOING TO HAPPEN?

By 2030, semiconductors will see transformative advancements, with 3nm and even sub-1nm chips becoming the norm, offering unparalleled processing power and energy efficiency. These innovations will enable more compact, faster, and intelligent devices across industries. Emerging technologies like quantum computing and photonics will begin to redefine the role of semiconductors, pushing computational limits and unlocking new possibilities in fields like encryption, AI, and scientific modelling.

Data centers will expand rapidly to meet the exploding demand for cloud computing, AI processing, and the proliferation of IoT devices, with their global energy consumption projected to rise to 8-10% of total output. However, sustainability will be at the forefront of their evolution. Data centers will increasingly adopt green energy solutions, such as solar, wind, and even nuclear power, along with advanced cooling systems like liquid immersion and cryogenic cooling to minimize their environmental footprint.

10,000 More Powerful GPUs

By 2030, GPUs (Graphics Processing Units) will be 10,000 times more potent than current models, dramatically improving the ability to handle complex computations for AI, deep learning, and simulations. These next-gen GPUs will deliver performance measured in petaflops and zettaflops, supporting tasks such as real-time AI inference, high-fidelity simulations, and rendering. These advancements in GPU power will enable breakthroughs in AI and scientific research, allowing computers to solve previously computationally impossible problems.

By 2030, semiconductor manufacturing will have progressed to the 0.5nm scale, continuing the trend toward more minor, more efficient chips. These nanometer-scale transistors will allow for greater processing power in smaller devices, pushing the limits of Moore's Law. Each shrink in transistor size will improve speed, energy efficiency, and processing capabilities, benefiting everything from consumer electronics to AI supercomputers. The shift from 5nm to 0.5nm will unlock new possibilities in quantum computing, advanced AI, and high-performance computing. In addition to making chips smaller, they will become three-dimensional in nature. 3D chips will become widespread, stacking layers of transistors vertically to improve performance and energy efficiency without increasing the chip's footprint. This architecture will enable data to move faster between layers, allowing for greater computational power and improved heat management. 3D chips will be essential in AI and big data applications, where dense data processing is required. These chips will revolutionize industries like AI, gaming, and autonomous systems by providing exponential performance boosts.

New materials such as graphene, carbon nanotubes, and nanomaterials will revolutionize chip manufacturing and electronics. These materials will allow for faster, more energy-efficient chips with improved heat dissipation, extending the limits of traditional silicon-based technologies. Quantum materials will also emerge, paving the way for quantum computing advancements. These new materials will contribute to breakthroughs in everything from quantum processors to more powerful energy storage systems.

Supercomputers will achieve computing power from several thousand petaflops to zettaflops (one zettaflop equals one billion petaflops). This level of computational power will enable real-time simulations of complex systems, from climate models to human brain simulations, and support advances in AI and space exploration. Zettaflop-capable systems will allow for unprecedented data processing, solving problems previously beyond reach.

Moore's Law predicts the doubling of transistors on a chip every two years and will face physical limits. Still, advancements in 3D chips, software, new materials, and quantum computing will extend its relevance and even accelerate the pace. With exponential progress in chip design and manufacturing techniques, the processing power of chips will effectively grow by a factor of 10. This acceleration will drive significant advances in AI, big data, and scientific research, enabling previously unimaginable computational feats.

Bio-computing could completely transform technology by using lab-grown brain cells, known as organoids, in place of traditional silicon chips. These tiny networks of living cells are incredibly fast and adaptable, processing information more efficiently while using only a fraction of the energy that current chips require. Unlike today's rigid hardware, organoids can learn and evolve, much like the human brain. Imagine a computing system no bigger than a pencil eraser yet powerful enough to outpace the fastest supercomputers while sipping on milliwatts of energy. This game-changing innovation could supercharge artificial intelligence, making it smarter and more dynamic, all while slashing the energy demands of data centers—one of the world's largest power guzzlers. By blending biology with technology, bio-computing might lead us into a future where computers are not only more intelligent but also far more sustainable.

AI to Design Better Chips to Make Better AI (Virtuous Cycle)

AI-driven chip design will create a virtuous cycle where AI systems are used to design better semiconductors, powering more advanced AI systems. This feedback loop will lead to increasingly powerful and efficient chips optimized for specific tasks such as machine learning, quantum computing, or autonomous systems. AI will analyze vast amounts of data on chip performance, identify patterns, and make real-time improvements to chip architectures, driving continual innovation in the tech industry. This virtuous cycle will accelerate technological progress across all fields.

AI models will distinguish between two core processes: training and inference. Training refers to teaching an AI model to learn from massive datasets, which will require vast computing power and time. On the other hand, inference is applying the trained model to make real-time decisions and predictions. Training will continue to consume most computational resources, but advanced GPUs, quantum computing, and energy-efficient chips will reduce the time required. Inference will be optimized for edge devices for faster, real-time applications, especially in healthcare, autonomous vehicles, and security.

Liquid cooling and immersion systems will be a mainstream solution in data centers and high-performance computing systems to manage the massive heat generated by increasingly powerful processors. Traditional air cooling will need to be improved for the dense, high-efficiency data centers required to support advancements like AI and quantum computing. Liquid cooling systems, which circulate coolant to absorb heat directly from the components, will significantly reduce energy costs and increase efficiency. Furthermore, the excess heat generated by data centers can be repurposed for growing cash crops like food or industrial hemp in nearby facilities. This process, known as waste heat recovery, will be a key part of sustainable data center design, contributing to local economies while reducing the environmental impact of energy consumption.

Excess heat from data centers will be increasingly used to support indoor farming and greenhouses. This waste heat can create the ideal growing environment for cash crops such as vegetables, fruits, or industrial hemp. Data centers, particularly those in colder climates, will partner with agricultural ventures to power year-round crop production, optimizing resources while minimizing waste. These greenhouses will also benefit from the stable heat source, helping offset energy costs and promoting sustainable, local food production. This synergy will play a significant role in cities' efforts to become more self-sufficient and environmentally responsible.

Cloud computing will be the backbone of digital infrastructure, allowing seamless access to processing power, storage, and software from anywhere. Cloud platforms will provide the flexibility to scale resources on demand, making them indispensable for businesses, AI development, and consumers. Cloud computing will integrate with edge computing to create hybrid systems, balancing centralized and decentralized processing to optimize speed, security, and efficiency. Companies and governments will rely on cloud-based platforms to store massive datasets and run AI-driven operations.

Meanwhile, edge computing will be crucial in decentralizing data processing, reducing latency, and easing the burden on massive centralized facilities. Edge computing will drive efficiency and responsiveness in critical applications by processing data closer to its source—whether in smart cities, autonomous vehicles, or industrial IoT systems. Together, these advancements will shape a digital ecosystem that is more powerful, sustainable, and adaptable than ever before.

WHY DOES IT MATTER?

Semiconductors and data centers are the backbone of our digital world, driving everything from essential online services to cutting-edge AI and 5G applications. As industries increasingly digitize, the ability to efficiently process and analyze vast amounts of data will be crucial for innovation and economic growth. With advancements in semiconductors, the evolution of technologies like AI, autonomous systems, and spatial computing would grow.

As the hubs of cloud computing and IoT ecosystems, data centers must scale to meet demand while addressing significant challenges, including data security, energy consumption, and environmental sustainability. Developing energy-efficient processors and renewable-powered centers will be critical to managing these concerns, especially as global reliance on cloud services grows.

Semiconductors will also play a transformative role in quantum computing, a technology poised to disrupt industries like healthcare, logistics, and cybersecurity by solving problems beyond the reach of classical computing. Their advancements will enable breakthroughs in personalized medicine, climate modelling, and advanced financial forecasting, making them indispensable to a connected, data-driven future.

Clean energy data centers (hydro, nuclear, hydrogen, other) will be essential to managing the increasing energy demands of cloud computing, AI, and other digital infrastructure. Renewable energy sources such as hydroelectric, solar, nuclear, and hydrogen power will dominate the energy supply of data centers, reducing their carbon footprint. Hydro-powered data centers will continue to thrive in regions with abundant water resources. In contrast, nuclear-powered centers will provide stable, high-output energy in areas seeking low-emission, reliable alternatives. Hydrogen fuel cells will also emerge as a clean energy source, supporting remote data centers and urban hubs. This shift to clean energy will help data centers become carbon-neutral, addressing the growing environmental concerns around energy consumption.

WHO WILL IT IMPACT?

The ripple effects of advancements in semiconductors and data centers will be profound, touching nearly every sector of society:

- **Tech Giants / Hyperscalers:** Companies like Google, Microsoft, and Amazon, which operate vast data center networks, will leverage advanced semiconductors to provide faster, more reliable cloud services. This will enable new capabilities in AI, big data analytics, and real-time applications for their users.

- **Hardware Manufacturers:** Semiconductor companies such as TSMC, Intel, and Samsung must innovate to meet demand for smaller, faster, and more energy-efficient chips, impacting everything from production processes to global supply chains.

- **Consumers:** Faster, more intelligent devices and services powered by cutting-edge chips will improve daily life, from seamless video streaming and instant translations to real-time gaming and immersive virtual experiences.

- **Automotive Industry:** The evolution of semiconductors will be pivotal in enabling autonomous vehicles, which require immense processing power for navigation, AI decision-making, and real-time communication with surrounding infrastructure.

- **Healthcare Sector:** Medical devices and AI-driven diagnostics will rely on advanced chips for precision and efficiency, revolutionizing treatments and patient outcomes through innovations like wearable health monitors and robotic surgery.

- **Telecommunications:** Semiconductors will support the global rollout of 6G networks, which will demand higher speeds, lower latency, and the ability to handle massive data volumes for IoT devices and next-gen communication systems.

- **Governments and Environmental Advocates:** With data centers projected to consume up to 10% of global electricity by 2030, policymakers and sustainability experts must address energy consumption through regulations, incentives for renewable energy adoption, and more brilliant cooling technologies.

From enabling real-time applications to solving complex global challenges, these advancements will drive the future of connectivity, efficiency, and innovation across all sectors.

CALL TO ACTION

The future of semiconductors and data centers will define the pace of innovation in the coming decade, but it also presents a challenge we can't afford to ignore. Businesses and governments must act now to meet the soaring demand for computational power without sacrificing sustainability. Start by investing in energy breakthroughs—think nuclear-powered data centers, advanced battery storage, or AI-driven cooling systems that make every watt count.

For businesses, reimagine your operations: integrate chips designed for maximum efficiency and support edge computing to reduce strain on centralized systems. Governments should go beyond generic policies and create actionable incentives—fund pilot projects for carbon-neutral data centers or partner with the private sector to accelerate sub-nanometer chip production that minimizes waste and energy use.

The tools to make this happen already exist; the challenge is scaling them before we hit critical limits on energy and resources. If we act boldly, we can build the infrastructure to support a brighter, faster, more sustainable digital future. The race is on—let's ensure we win it without losing the planet.

Cybersecurity and Privacy

Chester glanced out of his office window, the skyline a reminder of the vast digital frontier he helped protect daily. As the Chief Information Security Officer of one of the largest Fortune 100 companies, his job wasn't just about managing a team of cybersecurity professionals but about staying three steps ahead in a world where threats evolved by the second.

In 2030, the stakes were higher than ever. Quantum computers were nearing the capability to break conventional encryption, so Chester's team had been rolling out quantum-resistant encryption across all data networks. Every transaction and every sensitive file was now secured with lattice-based cryptography, a defence against the looming power of quantum decryption. Chester remembered when 256-bit encryption was the gold standard; now, it was as antiquated as a padlock on a safe.

Today was hectic. As he reviewed the latest security audit, his AI-driven monitoring system flagged a new phishing attempt. Unlike old phishing tactics, this one was sophisticated, AI-generated to perfectly mimic company executives' emails, using voice recognition deepfakes to leave audio messages. Chester quickly activated their advanced threat-response AI, which autonomously detected and isolated the threat, preventing any employee from opening fraudulent emails.

One of Chester's most significant initiatives was implementing a zero-trust architecture. In this system, every employee, device, and application had to verify itself continuously, no matter where it was or who was using it. It was the cornerstone of the company's defence strategy, ensuring no unauthorized access point could slip through the cracks.

The building's biometric security system flashed a quick notification as Chester walked through it, tracking his face and gait. Biometric authentication had become the default standard, replacing passwords entirely. Yet, Chester knew even this system wasn't foolproof; he had been running voice authentication tests to thwart the rise of deepfake technology, which could soon outwit even the most sophisticated biometric safeguards.

When Chester finally settled into his meeting, he was joined remotely by his team from around the world. They discussed potential vulnerabilities without fear of eavesdropping, thanks to quantum-encrypted communication channels. On Chester's tablet, his AI assistant displayed a map of the company's infrastructure, highlighting any new vulnerabilities in real-time, from Internet of Things (IoT) devices in warehouses to automated machinery in production lines.

As he wrapped up the day, Chester couldn't help but reflect on how his role had shifted. A decade ago, his focus was on basic firewalls and antivirus software. Now, he managed an intricate digital ecosystem fortified by quantum-resistant encryption, AI monitoring, blockchain-based identity management, and fully autonomous threat detection. The battlefront had changed, and Chester was prepared—equipped with the most advanced technology to secure not just a corporation but the confidence of millions of customers who relied on it.

CYBERSECURITY and PRIVACY IN 2030

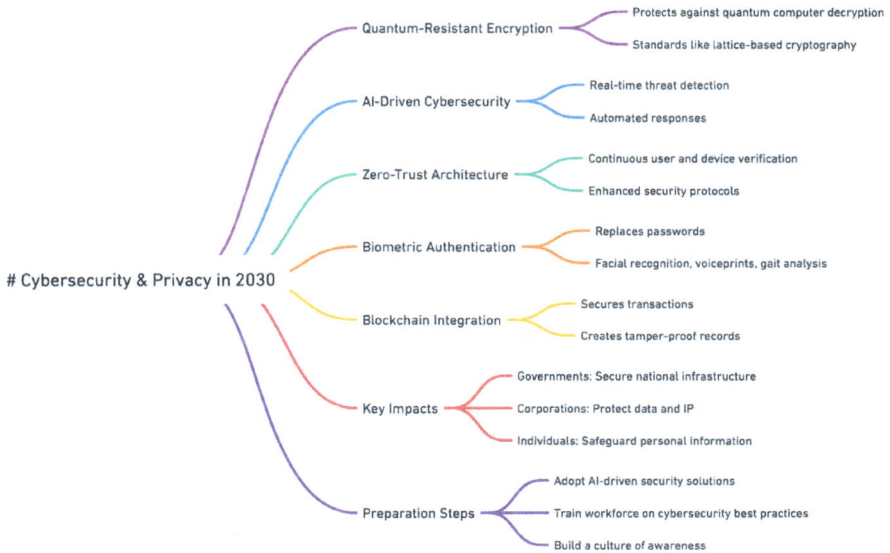

```
# Cybersecurity & Privacy in 2030
    ├── Quantum-Resistant Encryption
    │       ├── Protects against quantum computer decryption
    │       └── Standards like lattice-based cryptography
    ├── AI-Driven Cybersecurity
    │       ├── Real-time threat detection
    │       └── Automated responses
    ├── Zero-Trust Architecture
    │       ├── Continuous user and device verification
    │       └── Enhanced security protocols
    ├── Biometric Authentication
    │       ├── Replaces passwords
    │       └── Facial recognition, voiceprints, gait analysis
    ├── Blockchain Integration
    │       ├── Secures transactions
    │       └── Creates tamper-proof records
    ├── Key Impacts
    │       ├── Governments: Secure national infrastructure
    │       ├── Corporations: Protect data and IP
    │       └── Individuals: Safeguard personal information
    └── Preparation Steps
            ├── Adopt AI-driven security solutions
            ├── Train workforce on cybersecurity best practices
            └── Build a culture of awareness
```

PREDICTIONS

1. Quantum-resistant encryption will become the new standard.

2. AI-driven monitoring systems will autonomously detect and respond to sophisticated cyber threats in real time.

3. Zero-trust architecture will be widely implemented, requiring continuous verification of every user, device, and application.

4. Biometric authentication, including facial recognition, voiceprints, and gait analysis, will replace traditional passwords as the primary method of identity verification.

5. Blockchain technology will be extensively used to secure transactions and create tamper-proof records, particularly in critical sectors such as finance and government.

WHAT IS IT?

Cybersecurity refers to the practices, technologies, and processes designed to protect digital information, networks, and systems from unauthorized access, use, disclosure, disruption, modification, or destruction. In other words, cybersecurity is the art of safeguarding our digital lives from cyber threats, ensuring that our online activities remain private, secure, and uninterrupted. By 2030, cybersecurity will have evolved into an advanced, highly adaptive field where AI and automation are indispensable in protecting digital assets against a new generation of increasingly sophisticated threats. As cybercriminals leverage powerful AI algorithms and the vast computing potential of quantum machines, cybersecurity will incorporate cutting-edge technologies like quantum-resistant encryption, AI-powered threat detection, and decentralized security models such as blockchain, providing organizations with more robust, multi-layered defences.

Quantum-resistant encryption will become the backbone of secure communications, designed to withstand the decryption capabilities of quantum computers that can solve complex mathematical problems exponentially faster than classical systems. Traditional encryption standards, once thought unbreakable, will be replaced by lattice-based and post-quantum cryptographic algorithms that are theoretically resistant to quantum attacks, ensuring sensitive data remains secure even in the face of quantum computing power.

AI-powered threat detection will allow cybersecurity systems to operate in real time, constantly scanning and analyzing networks for anomalies and signs of intrusion. These AI systems will learn from vast amounts of data, adapting autonomously to new threats and identifying malicious patterns that would be impossible for human teams to recognize. Predictive algorithms will help organizations anticipate and neutralize cyber threats before they can exploit vulnerabilities, effectively shifting cybersecurity from a reactive to a proactive model. AI will also support the creation of intelligent firewalls and self-healing networks that can adapt to and counteract cyberattacks in milliseconds, further minimizing potential damage.

Decentralized security models, like blockchain, will protect digital identities and transaction records. Blockchain's immutable ledger will ensure transparency and security in data transactions, making it nearly impossible for attackers to alter or compromise critical records. This decentralized approach will allow cybersecurity systems to distribute control across networks, minimizing single points of failure and reducing the risk of large-scale breaches.

Biometric authentication will replace traditional passwords, adding an extra layer of security with multi-factor verifications that include facial recognition, voiceprints, and fingerprint scans. Users will no longer rely on passwords, which are vulnerable to breaches; instead, identity verification will be based on unique, personal attributes that are difficult to replicate or forge.

In addition, the **zero-trust architecture** will be standard practice, treating every user, device, and network interaction as a potential threat until it is verified. Continuous verification will become the norm, as will multi-factor biometric authentication systems that use a combination of facial recognition, voiceprint analysis, and fingerprint scans to grant access. With these tools, cybersecurity in 2030 will be far more resilient and adaptive, designed to address the evolving landscape of digital threats and the unprecedented speed and complexity of future cyberattacks.

What is Privacy? – The concept of privacy needs to be more understood and simplified. Privacy is not just about secrecy or hiding information. Instead, it is about controlling one's personal information and making informed decisions about its use. Privacy is the right to be left alone, to make choices about what information is shared, and to have some level of autonomy over one's data.

Privacy is not just about physical boundaries or spaces in the digital era. It concerns the digital footprint we leave behind, the data we share online, and the information we provide to companies, governments, and other organizations. Privacy has become more complex and multifaceted with the increasing reliance on digital technologies.

WHAT IS GOING TO HAPPEN?

By 2030, cybersecurity will be an advanced, AI-powered field focused on real-time protection. AI will instantly handle threat detection and automated responses, isolating threats before they spread. With quantum computing making traditional encryption outdated, quantum-resistant algorithms will protect sensitive data, especially in financial and government systems.

As 5G and IoT expand, increasing the number of connected devices, zero-trust architecture will be standard—continuously verifying every device and user to keep networks secure. Biometric authentication, like facial recognition and voiceprints, will replace passwords, adding personalized security.

Blockchain technology will secure transactions and data, creating tamper-proof records for finance and other critical sectors. Decentralized edge computing will enable faster, localized threat responses, making cybersecurity systems in 2030 highly adaptive, fast, and resilient against complex digital threats.

WHY DOES IT MATTER?

Cybersecurity is crucial in a highly digitized world where the impact of cyberattacks can be devastating. Effective cybersecurity protects against financial losses, data breaches, and infrastructure disruptions that could cripple essential services. As AI, robotics, and autonomous systems control more critical processes—from healthcare and finance to transportation and energy—securing these systems is vital to prevent large-scale damage, data theft, or even threats to national security. With personal and corporate data increasingly at risk, robust cybersecurity will be the backbone of trust and stability in a connected, automated society.

The economic impacts of cybercrime are staggering. According to a report by the Center for Strategic and International Studies (CSIS), the global cost of cybercrime is estimated to be around $6 trillion annually. This staggering figure is equivalent to the GDP of the world's 10th-largest economy.

For the average person, advanced cybersecurity means their data—like bank details, health records, and social media accounts—are protected against theft and misuse. Imagine a cyberattack that could wipe out a person's savings or leak sensitive health information; with strong cybersecurity, people can trust that their digital lives are secure, even as more personal data is shared across connected devices.

"For CEOs, cybersecurity means safeguarding the company's data, reputation, and operational continuity. The financial and reputational damage could be catastrophic if a cyberattack shut down production, accessed trade secrets, or disrupted supply chains. Strong cybersecurity enables CEOs to innovate and grow, knowing their company's assets, customer information, and competitive edge are protected, maintaining stakeholder trust and market stability.

WHO WILL IT IMPACT?

Cybersecurity advancements by 2030 will have wide-reaching impacts across multiple sectors:

- **Governments**

    - Will secure critical national infrastructure (power grids, transportation, healthcare systems) against cyberattacks that could disrupt public safety and essential services.

    - Will protect sensitive government data and defence systems from foreign threats, ensuring national security.

    - Will implement regulations to set high industry cybersecurity standards and protect citizens' digital rights.

- **Corporations**

    - Will safeguard proprietary data, trade secrets, and intellectual property, preventing theft that could harm competitive advantage.

- Will protect customer data and maintain trust, minimizing reputational risk in a cyber breach.

- Will invest in cybersecurity to ensure operational continuity, protecting automated systems and supply chains from disruptions that could cause financial loss.

- **Individuals**

  - Will experience excellent protection against identity theft and cybercrime, securing personal information across digital platforms.

  - Will benefit from enhanced privacy controls in a more connected world, with AI-powered tools that automatically monitor for suspicious activity on personal accounts.

  - Will gain confidence in using digital services (banking, healthcare, e-commerce), knowing personal data is well-protected.

- **Industries (Healthcare, Finance, Energy, etc.)**

  - The healthcare sector will rely on cybersecurity to protect patient records, secure medical devices, and prevent breaches that could compromise patient safety.

  - The finance industry will use advanced encryption and real-time fraud detection to protect against financial losses and preserve customer trust.

  - The energy sector will safeguard power grids and infrastructure, ensuring the reliability of essential services against cyber threats.

- **Digital Device and Service Users**

  - With more connected devices (IoT, smart home systems), individuals and businesses alike will benefit from cybersecurity systems that protect each device on the network.

  - As digital adoption grows, cybersecurity advancements will make digital services safer and more reliable for users worldwide.

CALL TO ACTION

To effectively prepare for the cybersecurity challenges of 2030, organizations must begin investing in AI-driven security solutions and upskilling their workforce in cybersecurity best practices.

Adopt AI-Driven Security Solutions: AI-powered cybersecurity tools can identify threats in real time, adapt to new attack patterns, and automatically respond to breaches, reducing reliance on human intervention. Autonomous threat response and self-healing networks will be crucial for organizations looking to stay ahead of increasingly sophisticated cyber threats, especially as quantum computing challenges existing encryption methods.

Prioritize Workforce Training in Cybersecurity: Employees are often the first line of defence; comprehensive training programs on cybersecurity best practices will equip them to recognize phishing, social engineering, and other attacks. Encourage a zero-trust mindset among the workforce, where verification is required at every level, reducing internal vulnerabilities and ensuring secure remote work practices. Organizations should establish regular cybersecurity drills, incident response simulations, and threat-awareness sessions to keep employees vigilant and up-to-date with the latest security protocols.

Build a Culture of Cybersecurity Awareness: Beyond technical training, fostering a culture where employees understand their role in securing company assets is essential for long-term resilience. Encourage cybersecurity responsibility across all departments, ensuring that cybersecurity isn't just an IT issue but a company-wide priority.

Types of Cyber Attacks

Cyber attacks can take many forms, and understanding the different types is crucial for developing effective defence strategies. The following are some of the most common types of cyber attacks:

- **Phishing Attacks:** Phishing attacks involve tricking individuals into revealing sensitive information by disguising it as a legitimate entity, such as login credentials or financial information. Phishing attacks can be emails, text messages, or social media messages.

- **Malware Attacks:** Malware attacks involve using malicious software to compromise a system or steal sensitive information. Malware can take the form of viruses, Trojans, or ransomware.

- **Denial of Service (DoS) Attacks:** DoS attacks involve overwhelming a system with traffic to make it unavailable to users. DoS attacks can disrupt business operations or extort victims' payments.

- **SQL Injection Attacks:** SQL injection attacks involve injecting malicious code into a website's database to extract sensitive information or disrupt system functionality.

- **Cross-Site Scripting (XSS) Attacks:** XSS attacks involve injecting malicious code into a website to steal sensitive information or take control of a user's session.

- **Ransomware Attacks:** Ransomware attacks involve encrypting a victim's files and demanding payment for the decryption key.

- **Social Engineering Attacks:** Social engineering attacks involve manipulating individuals into revealing sensitive information or performing specific actions that compromise system security.

- **Deepfake Attacks:** A deepfake attack involves the malicious use of AI-generated audio, video, or images that mimic real people with high accuracy, often to deceive, manipulate, or defraud by creating realistic but fake content that can bypass security measures, damage reputations, or spread misinformation.

WHAT HAPPENS IF QUANTUM SMASHES OUR CURRENT 356-BIT ENCRYPTION?

By 2030, quantum computing is expected to reach levels of computational power that could render current 356-bit encryption methods obsolete. If quantum computers successfully break these encryption standards, it would compromise the security of virtually all online systems—banking, communications, medical records, and more—exposing sensitive information to cybercriminals and state actors. This vulnerability could lead to widespread breaches, data theft, and loss of trust in digital systems. To mitigate this risk, urgent efforts to implement quantum-resistant encryption and upgrade legacy systems are crucial. Companies, governments, and individuals must prepare well for this potential security breach.

Quantum-Resistant Encryption

Quantum-resistant encryption will become the new gold standard in response to the quantum threat by 2030. These encryption algorithms will be designed to withstand the immense processing power of quantum computers, ensuring that even if quantum devices reach full capability, they will not be able to crack sensitive data. Technologies such as lattice-based cryptography, hash-based signatures, and multivariate quadratic equations are likely candidates for this new wave of encryption. Implementing quantum-resistant encryption will be essential to maintaining security for financial systems, government communications, and personal data, ensuring a safe digital landscape despite quantum advancements.

To safeguard your digital life, follow these best practices:

- **Strong Passwords:** Use unique, complex passwords and consider password managers.

- **Two-Factor Authentication:** Enable 2FA for added security.

- **Regular Software Updates:** Keep operating systems, browsers, and software up-to-date.

- **Secure Browsing:** Use HTTPS, avoid suspicious links, and be cautious with downloads.

- **Monitor Your Credit:** Regularly check credit reports and scores.

- **Be Cautious with Personal Information:** Limit sharing and be mindful of online interactions.

- **Travel:** Use public Wi-Fi cautiously, and consider VPNs for secure browsing.

- **Public Computers:** Avoid using public computers for sensitive activities.

- **Social Media:** Be cautious with personal information and limit sharing.

- **Online Shopping:** Verify websites, use secure payment methods, and monitor transactions.

Protecting our personal data, identities, and online safety requires a proactive approach. Individuals can take control of their digital lives by understanding the risks, consequences, and best practices outlined in this chapter. Remember, online safety is a shared responsibility, and by working together, we can create a safer, more secure digital landscape for all.

Material Science, 3D Scanning, and Printing

In 2030, the world had evolved into a vibrant tapestry of innovation, where technology reshaped lives profoundly. Among the many stories of transformation were those of Rahul, an 18-year-old in California, and Aaliyah, a pioneering scientist in the United Arab Emirates. Though separated by continents, their work would converge in a way neither could have foreseen.

Rahul's garage was a hub of activity. Rows of 3D printers worked tirelessly, crafting prosthetic limbs for individuals in war-torn regions. His passion for 3D printing had begun as a hobby, but it quickly transformed into a mission to provide affordable, custom solutions to those in need. The limbs he produced were made from advanced biopolymers—lightweight, durable, and biodegradable—designed for recipients in diverse environments, from mountainous regions to urban centers.

As the printers hummed, Rahul adjusted a holographic model of a prosthetic hand, ensuring its grip strength matched the needs of a young boy in Syria. "Nova," he said to his AI assistant, "run a durability test on this joint." Nova displayed a simulation in moments, showing the hand in action under various conditions. Satisfied, Rahul prepared the shipment for delivery, knowing that each limb represented more than mobility—it restored dignity and independence.

Meanwhile, thousands of miles away in Sharjah, Aaliyah was deep in her lab at the Sharjah Research Park. Her research focused on developing room-temperature superconductors, a breakthrough that promised to revolutionize energy systems worldwide. The lab was a marvel of modern science, equipped with AI-powered microscopes and quantum simulators capable of testing thousands of material combinations in mere hours.

Today, Aaliyah was analyzing a new hybrid material, an alloy of graphene and bio-synthesized compounds. The initial results showed promising conductivity levels that could enable lossless power grids and ultra-efficient devices. If successful, her work could power everything from electric vehicles to prosthetics, making energy more accessible and sustainable. She thought of the countless applications for her discovery and felt a spark of pride, knowing it could touch lives far beyond her lab.

The connection between Rahul and Aaliyah began with a simple article. Aaliyah stumbled upon an online feature about Rahul's work with 3D-printed prosthetics and was immediately struck by his ingenuity. Inspired, she reached out to him with an idea. "Your prosthetics could benefit from my material," she wrote. "It's lightweight and can harness ambient energy, eliminating the need for bulky batteries. Let's collaborate."

When Rahul received the message, he was stunned. The idea of integrating energy-harvesting materials into his designs was something he had dreamed of but needed more expertise to implement. Through the immersive metaverse, they connected, meeting in a shared virtual space that brought Rahul's garage and Aaliyah's lab together in a surreal yet seamless environment.

Their first conversation was electric. Aaliyah explained how her superconductive material could revolutionize the power supply for Rahul's prosthetics, allowing them to function indefinitely without charge. Rahul shared his designs, showing how the material could be adapted to his 3D-printed limbs. They brainstormed, sketched, and simulated, building on each other's ideas with an ease that transcended their physical distance.

As the partnership unfolded, Rahul integrated Aaliyah's material into his prosthetic designs. The new limbs were lighter, more efficient, and required no external power source—a game-changer for recipients living in remote areas. Meanwhile, Aaliyah's research gained a new dimension as she saw her material applied to real-world problems, fueling her drive to push boundaries further.

Despite their busy schedules, they stayed connected, often sharing updates on their progress. Rahul would send videos of children using the prosthetics, and their joy would be a testament to the impact of their work. Aaliyah, in turn, shared breakthroughs from her lab, showing how their collaboration inspired innovations in her field.

Their story was a microcosm of the transformative potential of 2030. Technology has empowered them to achieve their goals and connected them in ways that amplify their impact. Together, they embodied the promise of a new era—where innovation, empathy, and collaboration could unlock boundless possibilities for humanity.

MATERIAL SCIENCE, 3D SCANNING, and PRINTING IN 2030

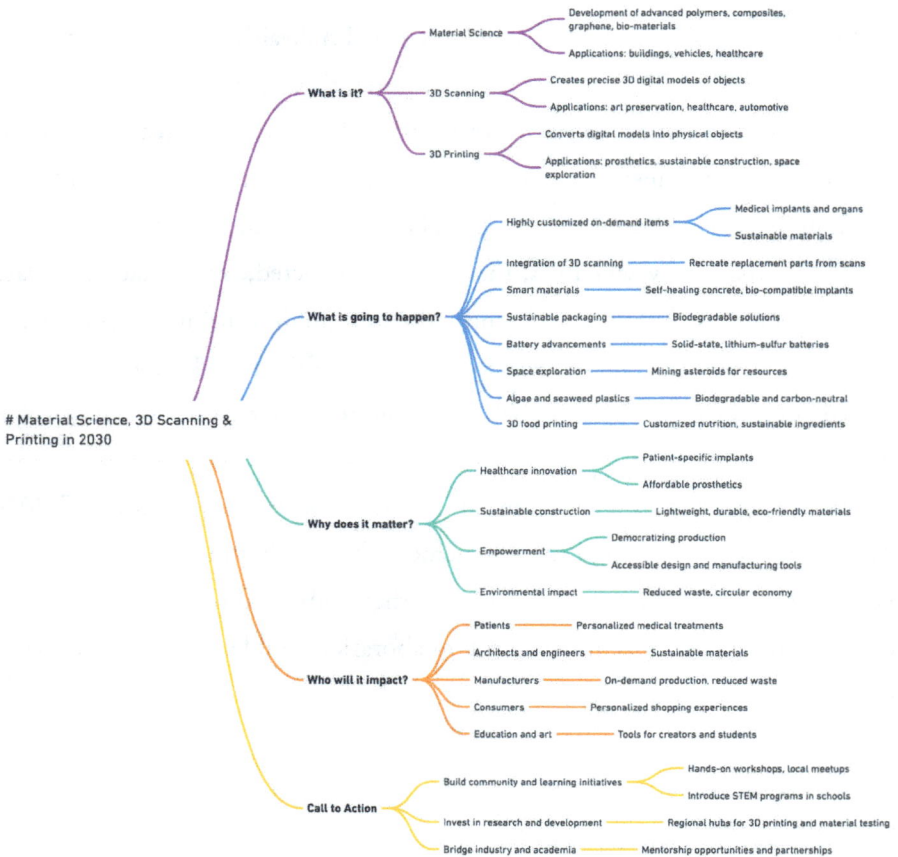

Material Science, 3D Scanning & Printing in 2030

What is it?
- Material Science
  - Development of advanced polymers, composites, graphene, bio-materials
  - Applications: buildings, vehicles, healthcare
- 3D Scanning
  - Creates precise 3D digital models of objects
  - Applications: art preservation, healthcare, automotive
- 3D Printing
  - Converts digital models into physical objects
  - Applications: prosthetics, sustainable construction, space exploration

What is going to happen?
- Highly customized on-demand items
  - Medical implants and organs
  - Sustainable materials
- Integration of 3D scanning — Recreate replacement parts from scans
- Smart materials — Self-healing concrete, bio-compatible implants
- Sustainable packaging — Biodegradable solutions
- Battery advancements — Solid-state, lithium-sulfur batteries
- Space exploration — Mining asteroids for resources
- Algae and seaweed plastics — Biodegradable and carbon-neutral
- 3D food printing — Customized nutrition, sustainable ingredients

Why does it matter?
- Healthcare innovation
  - Patient-specific implants
  - Affordable prosthetics
- Sustainable construction — Lightweight, durable, eco-friendly materials
- Empowerment
  - Democratizing production
  - Accessible design and manufacturing tools
- Environmental impact — Reduced waste, circular economy

Who will it impact?
- Patients — Personalized medical treatments
- Architects and engineers — Sustainable materials
- Manufacturers — On-demand production, reduced waste
- Consumers — Personalized shopping experiences
- Education and art — Tools for creators and students

Call to Action
- Build community and learning initiatives
  - Hands-on workshops, local meetups
  - Introduce STEM programs in schools
- Invest in research and development — Regional hubs for 3D printing and material testing
- Bridge industry and academia — Mentorship opportunities and partnerships

WHAT IS IT?

Material science, 3D scanning, and 3D printing are reshaping the world as we know it, unlocking possibilities that were once the stuff of science fiction. Material science, at its core, is the study of how we create and manipulate the building blocks of our world. It's behind the strength of the skyscrapers we build, the lightweight efficiency of electric cars, and even the biocompatible implants saving lives in hospitals. Over the years, breakthroughs in this field have led to the development of advanced polymers, composites, graphene, and bio-materials—innovations fuel the cutting-edge capabilities of technologies such as 3D printing.

3D scanning bridges the physical and digital worlds. Using sensors and advanced software, it captures every detail of an object—its shape, texture, and dimensions—to create a precise 3D digital model. Imagine being able to scan an ancient artifact to preserve it digitally or replicate a broken car part perfectly without the need for a physical mould. 3D scanning is already revolutionizing industries such as healthcare, automotive, and art preservation, allowing unparalleled precision and customization.

3D printing, or additive manufacturing, takes the process further, turning digital models into physical objects. Picture this: a blueprint on your computer becomes a tangible item built layer by layer from materials like plastic, metal, or ceramics. Need a custom prosthetic limb? A piece of architecture? Even a bio-printed organ? 3D printing makes it possible, combining efficiency with creativity. It is transforming everything from personalized medicine and sustainable construction to space exploration.

Together, these technologies form a powerful trio. Material science provides advanced building blocks, whether lightweight composites for aerospace or biodegradable plastics for consumer goods. 3D scanning creates the perfect digital models, and 3D printing turns those models into reality, pushing the boundaries of what we can design, build, and create. These tools will redefine industries in the coming decade, making production faster, more sustainable, and infinitely more innovative.

WHAT IS GOING TO HAPPEN?

3D printing and material science will be so advanced that creating highly complex, customized items on-demand will feel as routine as ordering online today. Imagine being able to print a custom medical implant perfectly tailored to your body, a functional human organ for a transplant, or sustainable building materials designed to withstand extreme climates—all with speed, precision, and minimal waste. This kind of manufacturing won't just be revolutionary; it will save lives, reduce costs, and create solutions that were unimaginable just a decade ago.

3D scanning technology will make this more seamless by allowing us to create exact digital replicas of physical objects. Do you need a replacement part for a car or a missing piece for an antique? Scan it, and a printer can recreate it in the tiniest detail. This will streamline industries like healthcare, where custom prosthetics and surgical guides will be printed from patient scans, and fashion, where garments will be tailored to exact body measurements. In automotive and aerospace, replacement parts will no longer need to sit in storage—they can simply be printed when required. You can try an early version of this using your iPhone (which now has an infrared LiDAR scanner built-in).

Meanwhile, breakthroughs in material science will introduce a new wave of innovative materials that adapt and respond to their environment. Think of self-healing concrete that repairs cracks on its own, dramatically extending the lifespan of roads and buildings or bio-compatible materials that seamlessly integrate with human tissue to help regrow skin or bones. These innovations will make construction, healthcare, and manufacturing not just more efficient but also more sustainable.

Algae and seaweed will be at the forefront of sustainable plastic alternatives, driving the shift away from petrochemical-based plastics. Researchers and companies are increasingly turning to these abundant, fast-growing marine plants as a source for bio-based plastics that can serve a wide range of uses, from packaging to textiles. Algae and seaweed-derived plastics are naturally biodegradable, non-toxic, and carbon-neutral, making them a key solution to reducing plastic pollution. These innovations will also have the potential to offset carbon emissions, as seaweed farms actively sequester carbon during growth, contributing to climate change mitigation.

In parallel, solutions for removing microplastics from oceans, water supplies, and the human body will gain momentum. Advances in nanotechnology and filtration systems will allow for more effective cleanup of these pervasive pollutants, helping restore ecosystems and improve human health.

Material science is also driving the development of more efficient and longer-lasting batteries, which will power everything from electric vehicles to smartphones, thanks to breakthroughs in solid-state batteries, lithium-sulfur, and other cutting-edge energy storage technologies. These batteries will charge faster, last longer, and be more environmentally responsible, supporting the transition to renewable energy sources.

Space exploration will take resource acquisition to a new frontier by mining asteroids. By 2030, missions will be in place to extract valuable materials, such as rare earth elements, platinum, and water, from asteroids. These resources will supplement Earth's supply of critical materials and support deeper space exploration and the potential for off-world habitats.

3D food printing will transform how we produce, prepare, and consume food, merging technology with culinary creativity. 3D food printers can create intricate, customized dishes layer by layer using edible ingredients such as purees, powders, and pastes. This technology will allow for precise control over food's nutritional content, texture, and appearance, revolutionizing industries from fine dining to healthcare. The ability to customize meals based on individual dietary needs will be particularly impactful for hospitals, elder care, and personalized nutrition plans, ensuring that patients and consumers get the exact nutrients they require.

Sustainability will be a key driver of 3D food printing. By using plant-based materials like algae, insect protein, or lab-grown meats, this technology will reduce food waste and lower the environmental impact of traditional agriculture. Additionally, 3D printing can use alternative proteins and ingredients to create dishes that mimic traditional foods, supporting global efforts to feed a growing population sustainably.

Restaurants and home kitchens alike will see benefits from this technology. Chefs will experiment with new textures, shapes, and flavours that were previously impossible, elevating the dining experience. Home users will have access to compact food printers that allow for creating gourmet meals at the push of a button or for making personalized, nutrient-rich snacks.

Even the way we think about resources will change. Using recycled and renewable materials such as biodegradable polymers and hempcrete and adopting additive manufacturing methods that generate almost no waste, we'll take giant steps toward a circular economy where products are designed to be reused and regenerated.

This isn't just about improving technology—it's about reshaping how we live and solve problems. By 2030, the tools at our disposal will make it easier to address global challenges, from healthcare access to sustainable housing, while opening up opportunities to innovate in ways we've never seen before. The future of material science, 3D scanning, and printing isn't just about what we can make—it's about how we'll use these capabilities to build a better, brighter world.

WHY DOES IT MATTER?

The convergence of material science, 3D scanning, and printing isn't just about new technologies—it's about changing how we create, build, and heal, impacting nearly every aspect of our lives. Traditional manufacturing methods, often wasteful and rigid, will give way to additive manufacturing processes that are far more efficient. These technologies will minimize material waste, cut costs, and allow customized designs that meet individual needs rather than rely on mass production.

In healthcare, this means life-changing innovations. Imagine a surgeon 3D-printing a patient-specific implant just hours before surgery or producing prosthetics and tissues tailored to a person's anatomy. These advancements will improve patient outcomes dramatically, making healthcare more precise, accessible, and affordable.

In construction, we'll see lighter, stronger, and more sustainable materials, reducing the environmental toll of building infrastructure. These breakthroughs will allow us to build faster and more intelligently, ensuring durability and sustainability in every project.

What's equally exciting is the democratization of production. Thanks to these advancements, individuals and small businesses can design and produce their own goods without needing massive manufacturing facilities. Whether a small entrepreneur creating a unique product line or a student building something that solves a community problem, these tools will empower more people to innovate and create.

Ultimately, this convergence will redefine industries, reduce environmental impact, and unlock unprecedented possibilities for creativity, innovation, and problem-solving.

WHO WILL IT IMPACT?

The impact of these technologies will be profound and far-reaching, touching nearly every corner of the globe and transforming industries such as healthcare, architecture, manufacturing, and retail.

Patients will benefit enormously as personalized medical treatments become the norm. Architects and engineers will gain access to sustainable, resilient materials, such as self-healing concrete and lightweight composites, enabling them to design structures that are not only cost-effective but also better equipped to withstand environmental challenges. Manufacturers will see drastic reductions in production times and costs, with additive manufacturing allowing for on-demand production and minimal waste. Consumers will experience a new era of personalization. Everything from custom-fit footwear and bespoke furniture to tailored clothing and accessories will become widely available, transforming shopping into a deeply individualized experience.

These technologies will also impact education, logistics, and even art, offering tools for creators, professionals, and students alike to innovate in ways never before possible. In short, this technological revolution will impact everyone—from industries to individuals—reshaping how we live, work, and interact with the world.

CALL TO ACTION

To fully embrace the transformative potential of material science, 3D scanning, and printing, we need to act now—and it starts with building a community and fostering learning. Companies and educational institutions should organize practical, hands-on workshops and training programs where people can explore these technologies, from engineers and designers to healthcare professionals and architects. These can take the form of weekly meetups at schools, libraries, or local community centers, creating accessible spaces for learning and collaboration.

For individuals, consider forming a local social group around 3D printing or material science, where members can share ideas, showcase projects, and brainstorm solutions to real-world challenges. Churches, schools, or maker spaces can host these gatherings, turning technology into a community-driven effort.

On a broader scale, governments should invest in research and development, funding breakthroughs and creating regional hubs for 3D printing and material testing facilities. These hubs could serve as accessible centers where small businesses, entrepreneurs, and students can experiment with cutting-edge tools without needing large-scale investment.

Finally, bridging the gap between industry, academia, and local communities will be key. Encourage schools to introduce STEM programs focused on 3D printing and material science—partner with businesses to create mentorship opportunities where professionals guide students and hobbyists in practical applications.

This isn't just about preparing for the future—it's about creating a culture of innovation today, one neighbourhood, school, or workplace at a time. By taking small, tangible steps now, we can ensure everyone is ready to thrive in a world redefined by material science, 3D scanning, and printing.

Construction, Housing, and Architecture

Chester, a world-renowned architect, is known for pushing the boundaries of sustainable, futuristic design. His latest project, a high-rise eco-hub for an urban neighbourhood, combines 3D-printed materials, robotics, and AI-driven planning systems to create a structure that houses people and actively contributes to environmental regeneration. His vision is brought to life by technology and skilled hands, and Zach, a veteran construction worker, is one of the key players on the ground.

Chester begins his day collaborating with his AI design assistant, which analyzes environmental data, structural demands, and energy requirements to suggest materials and layouts to optimize sustainability. The AI proposes using carbon-capturing concrete and recycled graphene composites to reduce the building's carbon footprint. Chester reviews and fine-tunes the AI's suggestions, enjoying AI's creative flexibility—ideas that once took weeks of planning and revisions are now efficiently curated and ready to implement. The AI also syncs with a building information modeling (BIM) system that updates in real-time, allowing Chester to visualize every aspect of the building's lifecycle, from foundation to final touches.

Meanwhile, Zach is on-site, working alongside robotic arms and autonomous machinery that handle heavy lifting and repetitive tasks such as bricklaying and foundation pouring. He supervises a bricklaying robot, ensuring its precision as it layers environmentally responsible bricks in intricate patterns that would have been time-consuming by hand. Drones fly overhead, surveying the site and providing Chester with real-time updates on progress and safety. Zach feels a sense of accomplishment in this new setup. While the robots take on physically demanding tasks, he can focus on more intricate work, refining details and solving problems that require human intuition. Integrating these robots has reduced construction time by nearly half and minimized risks in hazardous areas.

As the building rises, Chester and Zach connect through augmented reality (AR) headsets, which allow them to visualize the design elements in real-time. Chester can walk through a virtual version of the structure from his studio, overlaying instructions and tweaks directly onto Zach's AR display on-site. This seamless flow of communication and design enhances efficiency and reduces errors, creating a harmony between vision and execution. Prefabricated components support the building's modular design, assembled off-site with high precision and then transported by autonomous vehicles to be installed in record time, cutting costs and reducing environmental impact.

By evening, Chester reviews the day's progress and data from sensors embedded in the building materials. The AI system detects minor deviations in the temperature regulation of the concrete, instantly alerting Zach's on-site team to adjust settings for optimal curing. Thanks to these innovative materials and IoT sensors, Chester and Zach can ensure the building meets the highest sustainability and durability standards.

For Chester and Zach, technology has transformed their work, allowing them to focus on what they each do best—Chester on visionary design and Zach on skilled craftsmanship. As they near completion, the project stands as a testament to what's possible when human expertise and cutting-edge technology come together, creating structures that are as sustainable as they are groundbreaking.

CONSTRUCTION, HOUSING, and ARCHITECTURE IN 2030

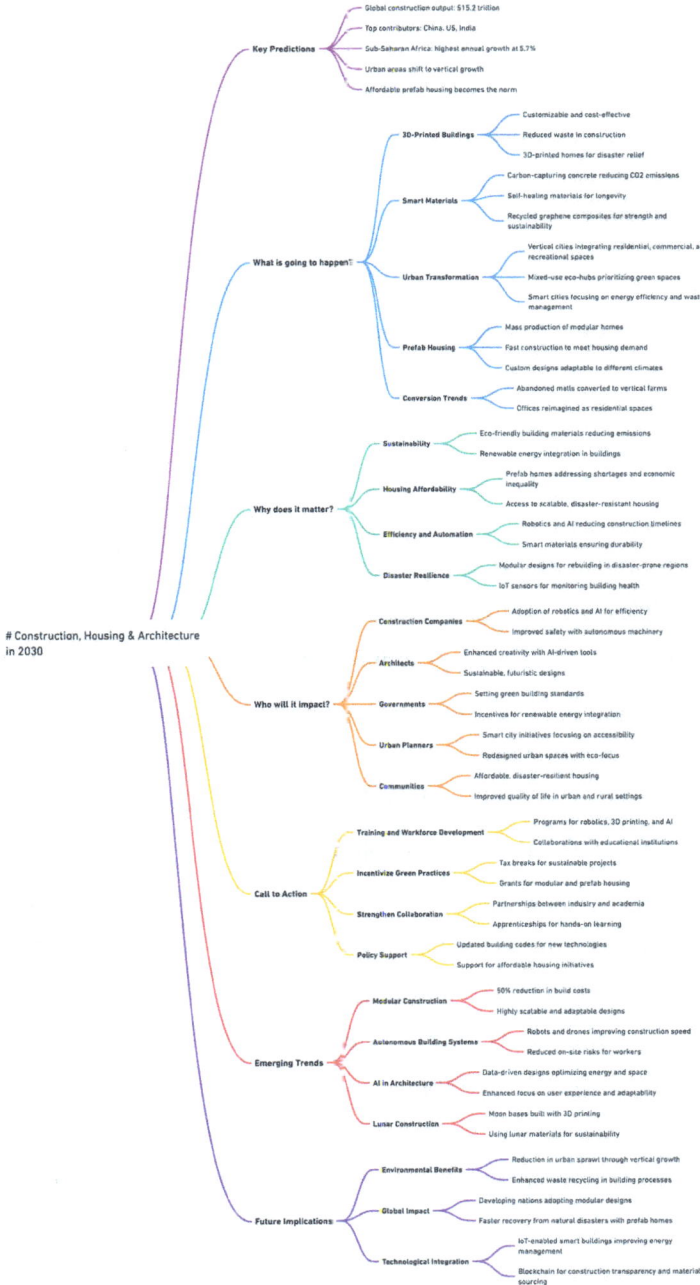

Key Predictions
- Global construction output: $15.2 trillion
- Top contributors: China, US, India
- Sub-Saharan Africa: highest annual growth at 5.7%
- Urban areas shift to vertical growth
- Affordable prefab housing becomes the norm

What is going to happen?
- 3D-Printed Buildings
  - Customizable and cost-effective
  - Reduced waste in construction
  - 3D-printed homes for disaster relief
- Smart Materials
  - Carbon-capturing concrete reducing CO2 emissions
  - Self-healing materials for longevity
  - Recycled graphene composites for strength and sustainability
- Urban Transformation
  - Vertical cities integrating residential, commercial, and recreational spaces
  - Mixed-use eco-hubs prioritizing green spaces
  - Smart cities focusing on energy efficiency and waste management
- Prefab Housing
  - Mass production of modular homes
  - Fast construction to meet housing demand
  - Custom designs adaptable to different climates
- Conversion Trends
  - Abandoned malls converted to vertical farms
  - Offices reimagined as residential spaces

Why does it matter?
- Sustainability
  - Eco-friendly building materials reducing emissions
  - Renewable energy integration in buildings
- Housing Affordability
  - Prefab homes addressing shortages and economic inequality
  - Access to scalable, disaster-resistant housing
- Efficiency and Automation
  - Robotics and AI reducing construction timelines
  - Smart materials ensuring durability
- Disaster Resilience
  - Modular designs for rebuilding in disaster-prone regions
  - IoT sensors for monitoring building health

Who will it impact?
- Construction Companies
  - Adoption of robotics and AI for efficiency
  - Improved safety with autonomous machinery
- Architects
  - Enhanced creativity with AI-driven tools
  - Sustainable, futuristic designs
- Governments
  - Setting green building standards
  - Incentives for renewable energy integration
- Urban Planners
  - Smart city initiatives focusing on accessibility
  - Redesigned urban spaces with eco-focus
- Communities
  - Affordable, disaster-resilient housing
  - Improved quality of life in urban and rural settings

Call to Action
- Training and Workforce Development
  - Programs for robotics, 3D printing, and AI
  - Collaborations with educational institutions
- Incentivize Green Practices
  - Tax breaks for sustainable projects
  - Grants for modular and prefab housing
- Strengthen Collaboration
  - Partnerships between industry and academia
  - Apprenticeships for hands-on learning
- Policy Support
  - Updated building codes for new technologies
  - Support for affordable housing initiatives

Emerging Trends
- Modular Construction
  - 50% reduction in build costs
  - Highly scalable and adaptable designs
- Autonomous Building Systems
  - Robots and drones improving construction speed
  - Reduced on-site risks for workers
- AI in Architecture
  - Data-driven designs optimizing energy and space
  - Enhanced focus on user experience and adaptability
- Lunar Construction
  - Moon bases built with 3D printing
  - Using lunar materials for sustainability

Future Implications
- Environmental Benefits
  - Reduction in urban sprawl through vertical growth
  - Enhanced waste recycling in building processes
- Global Impact
  - Developing nations adopting modular designs
  - Faster recovery from natural disasters with prefab homes
- Technological Integration
  - IoT-enabled smart buildings improving energy management
  - Blockchain for construction transparency and material sourcing

\# Construction, Housing & Architecture in 2030

PREDICTIONS

Global construction output is forecast to reach $15.5 trillion by 2030 (Global Construction 2030)

- China, the US, and India are projected to account for 57% of all global construction growth by 2030

- The construction output in the Asia Pacific region will reach $7.4 trillion by 2030

- Sub-Saharan Africa is projected to have the highest regional annual growth at 5.7%

WHAT IS GOING TO HAPPEN?

By 2030, advanced technologies like 3D printing, innovative materials, and sustainable design principles will profoundly influence construction, housing, and architecture. Construction will become more automated and efficient, with robots and drones aiding in everything from building to site inspection. Smart buildings made from environmentally responsible adaptable materials will offer enhanced energy efficiency, while prefabricated modular homes and 3D-printed structures will enable faster, more cost-effective housing solutions.

3D printing will become mainstream, allowing for the rapid construction of customized homes and commercial buildings. Smart homes with AI systems will autonomously manage energy consumption, climate control, and security. Urban planning will focus on vertical cities and sustainable housing developments, minimizing the environmental impact of sprawling metropolitan areas. Modular and prefabricated housing will also rise, enabling faster and more scalable housing solutions to meet global demand, especially in regions facing housing shortages.

WHY DOES IT MATTER?

The transformation in construction and architecture by 2030 will tackle critical global issues such as sustainability, housing shortages, and resource efficiency. As populations increase and cities grow, traditional construction methods will be unable to keep pace with demand, particularly in rapidly urbanizing regions. To address these challenges, the industry will embrace environmentally responsible building materials such as carbon-capturing concrete, graphene composites, and recycled materials that significantly reduce greenhouse gas emissions. These materials, designed to capture or offset carbon during production and use, will make new construction more sustainable, directly contributing to global efforts to combat climate change.

In addition to greener materials, automation, robotics, and 3D printing advances will drastically lower construction costs and speed up project timelines. Autonomous machinery and robotics will handle heavy, repetitive tasks, while 3D printing technology will enable on-site production of modular components tailored to specific building needs. This reduces the environmental impact of transporting large materials and allows for customizable designs suited to different climates and locations. These technologies will be particularly impactful in developing regions, where they will help alleviate housing shortages by creating affordable, durable structures faster than traditional methods allow.

Modular and prefabricated housing, constructed off-site and assembled on location, will address urban housing demands by allowing quick scalability and reduced waste. These prefab units, built to high standards and easily adaptable, will improve access to affordable housing globally, especially in areas that face critical shortages. Moreover, as construction costs decrease due to automation, affordability will improve, making decent, sustainable housing accessible to more people.

By integrating these innovative technologies and materials, the construction industry will shift from being a significant emitter of greenhouse gases to a driver of sustainable growth, with the potential to reshape cities for a greener, more efficient future. This comprehensive shift in how we build will support urban expansion and environmental goals, paving the way for sustainable development and improved quality of life worldwide.

WHO WILL IT IMPACT?

The transformation of the construction industry by 2030 will have distinct, targeted impacts on many stakeholders, including construction companies, architects, urban planners, consumers, governments, and communities in dense or disaster-prone areas.

For construction companies, integrating robotics, 3D printing, and automated systems will streamline projects, enabling them to work faster and more safely. For example, autonomous machinery like brick-laying robots and drones for surveying will take on labor-intensive tasks, freeing up skilled labour for more specialized roles. This change will also mean construction companies can work on multiple projects simultaneously, making them more competitive and responsive to market demands, especially as housing needs grow.

Architects will gain unprecedented creative freedom with access to AI-driven design tools and innovative materials. These tools will allow them to experiment with layouts and structural designs that were previously costly or time-intensive, bringing visionary ideas to life with minimal delay. As the demand for sustainable and resilient structures rises, architects will increasingly work on vertical cities, multi-use buildings, and eco-integrated projects—reshaping urban landscapes to reflect a sustainable, future-focused aesthetic.

Governments will play a key role in supporting this transformation. By setting incentives and standards for eco-friendly practices, they can encourage using renewable materials and regulate new building technologies to ensure safety and accessibility. Policies could also include financial incentives for affordable, resilient housing solutions, such as prefabricated homes, to make climate-adaptive housing more widely available, particularly for communities prone to natural disasters.

Urban planners will take advantage of the flexibility offered by modular and prefabricated structures to develop adaptable urban areas that prioritize green spaces, climate resilience, and accessibility. For example, with fewer personal vehicles and more shared or public transit, planners can repurpose traditional parking spaces for green spaces or community centers, making cities more livable and connected.

Finally, communities in densely populated and disaster-prone areas will experience direct benefits. With access to durable, low-cost housing that can be quickly deployed or modified, residents in these regions will be better prepared for extreme weather or population shifts and have a higher quality of life due to new housing technologies and materials designed for resilience.

This collective shift across stakeholders will redefine how construction projects are conceived, managed, and experienced, building toward a world where quality housing, sustainability, and resilience are available for all.

CALL TO ACTION

To prepare for the upcoming transformation in construction, the industry, governments, and educational institutions must invest significantly in training, infrastructure, and sustainable practices to support emerging building technologies. Construction companies must establish specialized training programs for workers to operate and maintain advanced robotics, 3D printing systems, and AI-driven design tools. These skills will become essential as the sector approaches automated and modular construction methods. To stay competitive, companies adopting these technologies early will also benefit from in-house training programs that familiarize their workforce with environmentally responsible materials, ensuring quality control and sustainability across projects.

Governments will play a critical role by funding public infrastructure projects that incorporates green building technologies and establishing incentives for sustainable development. By offering tax incentives or grants for companies that prioritize energy-efficient construction and resilient designs, governments can encourage the early adoption of sustainable practices. Additionally, governments should consider updating building codes and regulatory frameworks to accommodate new materials, modular components, and autonomous construction processes, ensuring that new technologies can be used safely and effectively. This regulatory support will help streamline approval processes for projects using innovative methods, giving cities a competitive edge as they strive to meet housing and sustainability goals.

Educational institutions must adapt quickly to meet the growing demand for sustainable construction and automation expertise. Universities and technical schools could develop specialized courses in sustainable architecture, construction automation, and AI in building design to equip the next generation of professionals with the skills they'll need in a transformed industry. Partnerships between schools, construction firms, and tech companies could also foster apprenticeship programs, providing hands-on experience in new technologies while ensuring a steady pipeline of skilled talent for the future.

Embracing these innovations early will allow construction companies, governments, and educational institutions to lead in creating innovative, sustainable housing solutions that meet both climate and population needs. Investing in infrastructure, training, and policy adjustments will improve competitiveness and position companies and cities as pioneers in the rapidly evolving world of sustainable construction.

A DEEPER DIVE

3D Printed Houses, Buildings, and Skyscrapers Become Commonplace

By 2030, 3D printing will be a mainstream method for constructing homes, commercial buildings, and skyscrapers. Using advanced materials like concrete, polymers, and recycled composites, 3D printers will allow for rapid, cost-effective, and environmentally friendly construction. This will enable architects to design complex, custom structures at a fraction of the cost of traditional building methods, with the ability to build a house in a few days. 3D printing will revolutionize urban and rural construction, offering affordable housing solutions and drastically reducing construction waste and labour costs.

Cities Build Up Instead of Out

Urban areas will increasingly focus on vertical development, building up rather than out, in response to population growth and the need for sustainable city planning. Skyscrapers and vertical cities will become the norm, incorporating mixed-use spaces, including residential, commercial, and recreational areas, all in one structure. These vertical cities will prioritize energy efficiency, waste reduction, and green spaces. The shift toward building up will help preserve land, reduce urban sprawl, and make cities more livable by reducing traffic and pollution.

Rural Communities Start to Participate in the Global Economy

By 2030, rural communities will have more direct participation in the global economy, thanks to improved internet connectivity, automation, and remote work technologies. Access to high-speed broadband and advancements in telecommunications will allow individuals in rural areas to work for companies anywhere in the world. Additionally, local agriculture and small-scale manufacturing industries will benefit from new distribution networks, enabling rural economies to thrive and contribute to global supply chains. This shift will help bridge the economic divide between urban and rural areas.

Inexpensive Pre-fab Housing Allows Everyone to Own a Home

Prefab housing will become a widely available and affordable option for home-ownership by 2030. Advances in modular construction and 3D printing will make it possible to produce high-quality, energy-efficient homes at a fraction of the cost of traditional construction. Prefab homes will be customizable, built off-site, and assembled quickly, reducing labour costs and making homeownership accessible to a broader population. Governments and private sectors will promote these housing options to solve housing shortages and economic inequality.

Governments Prevent Large Corporations from Owning Residential Real Estate

In 2030, governments will implement policies to limit large corporations' ability to buy and control residential real estate. These measures will be enacted to combat rising housing costs and ensure that residential properties remain affordable for individuals and families. These regulations will focus on preventing real estate monopolies and encouraging homeownership rather than large-scale corporate investments in housing. This will help balance housing markets and ensure more people have access to homes rather than competing with corporate interests.

Conversion of Office and Commercial Real Estate to Residential

As remote work becomes the norm by 2030, many vacant office and commercial buildings will be repurposed as residential spaces. The shift from traditional office environments will lead to creative conversions of underutilized buildings into apartments, co-living spaces, and mixed-use developments. These conversions will breathe new life into urban centers, providing affordable housing options while reducing the environmental impact of new construction. This trend will also revitalize downtown areas, making them more vibrant and community-oriented.

Use of Old Malls for Local, Vertical Farms

By 2030, old, underused malls will be transformed into local vertical farms, contributing to urban food production. Vertical farming, which uses stacked layers to grow crops indoors, will take advantage of the ample space in abandoned retail buildings. These vertical farms will provide fresh, locally grown produce to urban populations, reducing transportation costs and environmental impact. The conversion of malls into agricultural hubs will support sustainable food systems, revitalize empty commercial spaces, and create jobs in urban farming.

Modular Designs Save 50% in Build Costs.

In 2030, modular construction will be a standard method for building homes and commercial spaces, saving up to 50% in construction costs. These modular designs are pre-manufactured in factories and assembled on-site, reducing labour costs, waste, and construction time. Modular construction is highly adaptable, allowing for easy customization and scalability, making it an attractive option for residential and commercial projects. This efficiency will help meet the growing demand for housing while addressing the challenges of affordability and sustainability.

Building on the Moon

By 2030, the concept of building on the moon will move from science fiction to reality. NASA and private companies like SpaceX will begin constructing the first moon bases using 3D printing and lunar materials. These bases will serve as research hubs and waystations for deeper space exploration, such as missions to Mars. The ability to build infrastructure on the moon will mark a significant milestone in humanity's expansion into space, creating opportunities for scientific discovery, resource extraction, and potential colonization.

AI Will Help Design New Buildings

AI will be critical in designing buildings by 2030, optimizing everything from structural integrity to energy efficiency. AI algorithms will be able to analyze data from various sources, such as environmental conditions, material properties, and usage patterns, to create building designs that are highly efficient and sustainable. These AI-generated designs will incorporate innovative technologies, reduce waste, and provide more creative and customized architectural solutions. Architects will collaborate with AI systems to push the boundaries of design innovation.

Autonomous Building Systems

By 2030, autonomous building systems will handle much of the construction process, from site preparation to finishing touches. Robots and AI will manage excavation, bricklaying, and painting tasks, significantly speeding up construction and reducing human labour costs. These systems will be capable of working 24/7, improving productivity and allowing projects to be completed in record time. Automation will also lead to safer construction sites by minimizing human involvement in hazardous tasks.

Robot Construction in 2030

By 2030, robot construction will be a game-changer across the entire building industry, with robots handling various tasks, from large-scale construction to fine-finishing work. These machines will perform specialized roles such as bricklaying, drywall installation, and welding, improving precision and reducing human error. For example, brick-laying robots can work faster than humans, ensuring uniformity in the placement of bricks, while drywall bots will automate the measuring, cutting, and installation of drywall sheets. Concrete-pouring robots will improve speed and consistency in laying foundations, while painting and finishing robots will deliver impeccable results without manual labour.

Robots will also excel in hazardous environments, such as high-rise construction or underground tunnelling, performing tasks that reduce the risk to human workers. Drones will survey construction sites, map out topography, and inspect structures for quality assurance. Combined with AI and autonomous systems, robots can work around the clock, accelerating construction timelines and lowering costs. The versatility of construction robots will result in more efficient, cost-effective, and safer building practices, revolutionizing how we build everything from homes to skyscrapers to infrastructure projects.

As robotics becomes more sophisticated, their integration into the construction industry will help address labour shortages, improve safety standards, and ensure higher quality and consistency in every project.

KEY FACTORS CONTRIBUTING TO THE TRANSFORMATION

- **Technological Advancements**: The widespread adoption of building information modelling (BIM), 3D printing, and other digital technologies has revolutionized the construction process, enabling greater precision, efficiency, and sustainability.

- **Sustainability and Environmental Concerns**: Growing awareness of climate change, energy efficiency, and environmental sustainability has led to a focus on green building practices, renewable energy, and environmentally responsible materials.

- **Changing Consumer Preferences:** Shifts in consumer behaviour, driven by demographic changes, urbanization, and the sharing economy, have led to increased demand for flexible, adaptable, and experiential spaces.

- **Regulatory and Policy Changes:** Evolving building codes, zoning regulations, and policy initiatives have created new opportunities for innovation and sustainability in construction and architecture.

- **Increased Focus on Sustainability:** The emphasis on environmental sustainability has led to the development of green building certifications, such as LEED and WELL, and the integrating of renewable energy systems.

- **Construction Management Software:** Construction management software is being used to streamline the construction process. This software enables architects, engineers, and contractors to collaborate more effectively, reducing errors and improving communication. Construction management software can also be used to track progress, manage schedules, and monitor budgets.

- **Shift to Experiential Design:** The focus on experiential design has led to the creation of immersive, engaging, and adaptive spaces that prioritize user experience and well-being.

- **Sustainable Building Materials:** Critical to the construction industry's shift towards sustainability, these materials are designed to reduce buildings' environmental impact while improving indoor air quality and occupant health. Sustainable building materials include recycled glass, bamboo, and low-VOC paints.

Synthetic Biology

Adrian adjusted the dim light in his lab, and the soft hum of bioreactors filled the room. He was on the verge of a breakthrough—a custom-designed synthetic virus capable of reprogramming cells to reverse aging at a molecular level. For years, Adrian had been driven by the dream of unlocking the body's innate potential to heal and renew itself, and now, in 2030, he stood at the cusp of making that dream a reality.

On the other side of the world, Max, a sprightly 110-year-old, jogged slowly but steadily along the beach. His step had the deliberate rhythm of someone decades younger. Max wasn't just alive—he was thriving. His vitality resulted from personalized medicine that had redefined what aging could mean. A synthetic biological protocol, developed in part by Adrian's team, continuously monitored and repaired Max's cells, ensuring his body operated at peak efficiency.

Adrian's work wasn't purely academic. His father had died young from a degenerative disease that today would be easily treatable. That loss fueled his relentless pursuit of synthetic biology's promise. "What if we could rewrite the script of life?"

he often asked his colleagues. And rewrite it he had—creating biological circuits that acted as molecular engineers, repairing DNA, eliminating senescent cells, and preventing age-related decline.

Max was one of Adrian's earliest patients. Ten years ago, he'd been an aging war veteran plagued by arthritis, heart disease, and fading memory. Today, he managed a community garden, mentoring younger generations on sustainable living, his mind as sharp as it had been in his youth. The secret wasn't just in the treatments— it was in the harmony between Max's adherence to his regimen and the precision of Adrian's synthetic biology protocols. Max's bloodstream hosted billions of bioengineered organisms designed to monitor, repair, and optimize.

Their worlds collided unexpectedly. One evening, Adrian received a video message from Max. In the footage, Max stood beside a young man who had been severely injured in a car accident. "This kid doesn't have the time to wait for approvals and protocols," Max said, his voice firm but pleading. "You have to help him. Do what you did for me."

Adrian felt the weight of the decision. The synthetic treatments were groundbreaking but still highly controlled due to ethical and regulatory concerns. Yet Max's life was proof of their potential. After hours of internal debate, Adrian decided to act. He initiated a call with Max and the young man's family, outlining a compassionate use program that could bring the boy into a trial. "This isn't just science," Adrian said. "It's about the lives we can save now, not in some distant future."

Weeks later, Adrian stood beside Max in the garden. "You were right to push me," Adrian admitted. "It's easy to lose sight of why we do this when you're bogged down in the details. But this work—it's for people like you and him."

Max smiled, the lines on his face a testament to the years lived, not the years lost. "We have the tools to rewrite our stories," he said. "But it's up to people like you to ensure we write the right ones."

Their conversation trailed off into the rustling of leaves and children's laughter nearby. Both knew their work was far from over. Still, for that moment, they allowed themselves a shared sense of purpose and hope—a future where the boundaries of life and health were no longer dictated by chance but by the transformative power of synthetic biology.

SYNTHETIC BIOLOGY IN 2030

WHAT IS IT?

Synthetic biology is not just a branch of science; it is a paradigm shift that merges biology and engineering, creating lifeforms with characteristics we design. By 2030, synthetic biology will redefine how we address health, agriculture, and environmental sustainability.

Synthetic biology, through the deliberate manipulation of living systems, empowers scientists to transcend traditional biological boundaries. By integrating engineering principles into living organisms, synthetic biology aims to harness the potential of biological mechanisms to create custom-designed biological circuits, novel genetic pathways, and entirely new lifeforms capable of performing tasks beyond what naturally evolved organisms can achieve.

WHAT IS GOING TO HAPPEN?

The transformative impact of synthetic biology in 2030 will be pervasive, impacting several areas of our lives.

Healthcare Revolution: The field will redefine disease management by creating synthetic organisms and engineered tissues that replace or repair damaged biological structures. For instance, we can produce bio-synthetic stem cells to repair or replace faulty tissues and treat chronic diseases in a highly personalized manner. Synthetic viruses and bacteria will be programmed to locate and eradicate cancer cells with pinpoint accuracy, reducing the need for toxic chemotherapy and significantly increasing survival rates.

Case Study: Maria's Personalized Cancer Treatment. Maria, a 45-year-old mother of two, was diagnosed with an aggressive form of breast cancer. Traditional chemotherapy poses high risks of severe side effects. However, with the advent of synthetic biology, an artificial virus tailored to her genetic makeup was injected to target and eradicate her cancer cells specifically. Within months, Maria was cancer-free without experiencing the devastating side effects of chemotherapy. For Maria, synthetic biology didn't just cure her cancer; it gave her a second chance at life and precious time with her children.

Agricultural Transformation: Synthetic biology will enable the development of drought-resistant, pest-resistant, and nutrient-enhanced crops. Innovations will help ensure food security and resilience in the face of climate change, and bio-engineered microorganisms will be used to enrich soil health, creating a sustainable agricultural cycle. Engineered nitrogen-fixing bacteria will allow non-leguminous crops to self-fertilize, reducing the reliance on chemical fertilizers and minimizing runoff pollution.

Case Study: Raj's Resilient Crops. Raj, a farmer in India, used to struggle with unpredictable monsoons and soil degradation that severely impacted his crop yield. With synthetic biology, Raj now uses seeds genetically engineered for drought resistance, and engineered bacteria are added to his fields to enrich the soil naturally. The results are higher crop yields and financial stability. For Raj, synthetic biology is not just about technology—it's about securing his family's future and empowering his community to thrive.

Environmental Sustainability: Bio-factories and engineered microorganisms will be deployed to reduce pollution and restore degraded ecosystems. Imagine synthetic algae that efficiently capture atmospheric carbon dioxide or engineered bacteria that digest plastic waste in oceans. New forms of biofuels will replace traditional fossil energy sources, offering cleaner and more sustainable alternatives. Synthetic biology will allow for the creation of specialized microbes capable of breaking down even the most persistent pollutants, such as heavy metals, contributing to the revival of contaminated environments.

Case Study: The Revival of Lake Pinewood. Once a thriving habitat, Lake Pinewood became heavily polluted with plastic waste and toxic chemicals due to years of industrial runoff. By 2030, synthetic biologists released engineered microorganisms into the lake, breaking down plastics and neutralizing toxins. Within a year, fish and plant life returned, and the lake again became a place where the local community gathered. Synthetic biology represented not only environmental restoration but also the restoration of a lost cultural space.

Synthetic biology transforms life, ushering in a world where the concept of "natural" expands to include the deliberate creations of human ingenuity.

A DEEPER DIVE

Synthetic biology is poised to revolutionize industries and redefine our lives, offering transformative solutions in healthcare, agriculture, and environmental restoration. However, with this immense potential comes the responsibility to steer these advancements toward equitable and ethical outcomes.

Synthetic biology is not just about altering DNA but about reshaping our relationship with life. By harnessing its capabilities, we can eradicate diseases, combat climate change, and create sustainable systems that benefit everyone. But the true power of this technology lies in our collective ability to wield it wisely.

We must embrace this moment with excitement and caution as individuals, communities, and policymakers. We have the tools to solve some of humanity's most significant challenges—but only if we prioritize collaboration, inclusivity, and rigorous ethical considerations.

PREDICTIONS FOR 2030

CRISPR Therapies: CRISPR technology will be refined into an everyday tool for precisely editing genes. By 2030, it will enable treatments for genetic disorders, cancers, and other diseases that were once untreatable. Beyond disease correction, CRISPR could boost immunity, regulate metabolism for obesity management, and even slow aging by repairing damaged DNA. New delivery mechanisms for CRISPR therapies will be available, such as synthetic viral vectors, nanoscale robots, or engineered proteins that improve the accuracy and safety of genetic interventions. Advances in CRISPR's precision will help minimize off-target effects, making it safer and more reliable for clinical applications.

Synthetic Organs and Tissues: Lab-grown organs will become a viable alternative to traditional transplants. Organs engineered from patient-specific cells will reduce rejection rates and eliminate waiting lists. Organoids—miniature versions of organs—will also be used to test the effectiveness of new drugs, dramatically cutting down development timelines. Tailored to the patient's immune system, synthetic lungs, and kidneys will lead to breakthroughs in treating chronic conditions such as Chronic Obstructive Pulmonary Disease and renal failure. Advances in scaffolding technologies and biomaterials will make it possible to grow entire limbs, potentially offering solutions to those with physical disabilities or victims of accidents. These synthetic tissues will be functional and include nerve integration, allowing for sensation and motor control comparable to natural limbs.

Biomanufacturing: Engineered organisms will be routinely used in industrial settings to produce chemicals, biofuels, pharmaceuticals, and biomaterials, effectively replacing petroleum-based production. Biomanufacturing will allow for the local, on-demand creation of complex molecules, making supply chains more resilient and less vulnerable to disruptions. New methods in metabolic engineering will enable microbes to synthesize everything from perfumes and food additives to therapeutic proteins and biodegradable plastics. The rise of bio-foundries—automated facilities that design, build, and test engineered biological systems—will significantly accelerate innovation and commercialization in synthetic biology.

PREDICTIONS FOR 2040

Full-Body Regeneration: Advances in stem cells and synthetic tissue technologies will lead to the regeneration of entire body parts or organs. Synthetic biology and bio-printing will converge to allow complete limb regeneration. Cancer survivors will benefit from regenerating damaged tissues, avoiding long-term disabilities. Artificial skin, complete with sweat glands and pigmentation, will also be possible, providing full functionality and aesthetic appearance for burn victims. New medical protocols will allow regeneration without surgery, using injections of self-assembling tissue and advanced gene circuits. This will include regenerating nerves, muscles, and bones, restoring complete functionality to individuals who suffer from degenerative diseases or traumatic injuries.

Bioengineered Ecosystems: Synthetic biology will create ecosystems capable of withstanding and reversing environmental damage. Engineered bacteria could break down persistent pollutants like heavy metals and plastics, while specially designed fungi could support biodiversity recovery by improving degraded soil quality. Coral reefs and other delicate ecosystems will benefit from engineered species specifically designed to survive higher temperatures and acidified waters, helping to maintain marine biodiversity. Synthetic organisms will help stabilize ecosystems by creating symbiotic relationships with native species, reducing vulnerability to climate change.

Personalized Genomic Editing: Beyond treatment, gene editing will allow enhancements such as increased muscle efficiency, reduced aging markers, or enhanced cognitive capabilities. The availability of precision genomic editing kits will give people unprecedented control over their health, potentially enabling individuals to optimize their immune systems against emerging pathogens or enhance their resistance to chronic diseases like diabetes. Gene-editing therapies will be personalized based on an individual's genomic, microbiome, and epigenetic profile, offering tailor-made solutions for optimizing health and preventing disease. This personalized approach will be available even before birth, as parents could choose to eliminate genetic susceptibilities for their children.

DOES IT MATTER?

- **Solve Critical Challenges:** Synthetic biology will solve global health issues, food insecurity, and environmental challenges. It promises to offer medical solutions for genetic diseases such as Huntington's and cystic fibrosis, conditions that currently have limited treatment options. Additionally, engineered bacteria could be deployed to areas suffering from pollution crises, providing an effective and natural solution for environmental clean-up.

- **Lower Healthcare Costs:** By shifting the focus from symptom management to disease correction and prevention, healthcare costs could drop significantly, improving the quality of life globally. Preventative gene editing will reduce the prevalence of chronic diseases, and biomanufacturing could produce life-saving drugs at a fraction of their current cost.

- **Food and Climate Security:** By 2030, bioengineered crops will ensure stable food supplies, even in areas facing extreme climate challenges. Engineered organisms will clean up oil spills, plastics, and other pollutants, making environmental restoration more effective. Plants designed to thrive on marginal lands will transform previously unusable territories into productive agricultural areas, ensuring food security for growing populations.

WHO WILL IT IMPACT?

1. Patients:

- Individuals suffering from genetic diseases, organ failure, or cancer will have new treatment opportunities that will radically alter their prognosis. They will receive personalized medical interventions based on real-time genetic assessments, improving outcomes and reducing side effects.

- People with disabilities will find new hope in regenerative options, as bioengineered organs and tissues can restore lost functions. Synthetic nerve grafts will reconnect damaged neural pathways, helping patients regain sensory and motor functions that were previously irrecoverable.

2. Healthcare Systems:

- Health systems must evolve to incorporate synthetic organ transplants, bioengineered treatments, and precision medicine. Regulatory frameworks will be updated to accommodate the ethics and safety of these new interventions. New medical professions specializing in synthetic biology and genomic editing will emerge, bridging the gap between biological sciences and clinical practice.

- Insurance models will shift towards preventive and regenerative treatments rather than costly long-term care. Value-based care approaches will become more prominent, where insurance covers gene editing and biomanufacturing treatments that offer long-term health benefits.

3. Agriculture and Environmental Sectors:

- Farmers will benefit from bioengineered crops that are more resistant to pests, reduce the need for chemical pesticides, and improve yields. These genetically optimized crops will also be designed to produce more nutritional content, reducing the reliance on supplements in malnourished populations.

- Conservationists and environmental organizations will use engineered organisms to restore degraded habitats, reduce pollution, and tackle climate change. Microbial consortia will be used in reforestation projects to enhance plant growth and resilience, ensuring restored ecosystems can withstand the pressures of a changing climate.

CALL TO ACTION

Genetic Literacy: Start now by understanding the basics of synthetic biology. The coming decades will bring revolutionary changes that impact health, agriculture, and the environment. Learning how gene editing, biomanufacturing, and synthetic ecosystems work will prepare you to make informed decisions, participate in discussions, and benefit from these changes. Engage with resources like online courses, podcasts, and books on synthetic biology to develop foundational knowledge that will help you navigate this fast-evolving field.

AI, Robotics, and the Future of Work

Marti awoke with the hum of machinery echoing softly in the vast, bright factory. Rows of machines and tools gleamed, each in perfect order, as the day's work began. Marti's task? Building the intricate machinery that powered the city's autonomous vehicles. Precision was everything, and every component had to be flawless.

Marti moved gracefully between workstations, guided by an artificial inteligence (AI) assistant that optimized workflows in real-time. Each task was planned for maximum efficiency: seamlessly assembling, welding, and testing components. The ever-watchful AI provided alerts for maintenance needs or flagged anomalies in production lines. Marti felt a sense of accomplishment knowing that each creation contributed to a safer, cleaner transportation future.

Meanwhile, Vivek started their day in a serene suburban home surrounded by lush greenery. Vivek's responsibilities were centred on maintaining the household—a job that, once deemed mundane, had been reimagined for the modern era. The yard was meticulously tended with tools adjusted based on weather data.

Inside, the floors gleamed as cleaning systems quietly hummed under Vivek's direction, ensuring no dust was missed.

Laundry, often considered a tedious chore, became a masterpiece of efficiency in Vivek's hands. Sensors analyzed each fabric's needs, recommending precise cycles, while AI folded and organized garments with precision. The AI assistant, connected to the home's ecosystem, alerted Vivek when the fridge needed restocking or energy usage could be optimized further. Vivek moved purposefully, each action contributing to a well-functioning home environment.

The worlds of Marti and Vivek collided one crisp autumn morning at a community hub, where creators, innovators, and doers gathered to share ideas. Fresh from solving a manufacturing bottleneck, Marti spoke passionately about the need for precision and efficiency in industrial systems. Ever-focused on sustainability, Vivek shared insights on how homes could become more self-reliant, reduce energy waste, and optimize daily life.

As their discussion deepened, they realized their shared mission: advancing productivity and harmony in a technology-driven world. They envisioned a future where factories and homes coexisted as symbiotic systems—where innovation in one sphere could inform and elevate the other. The conversation ended with a mutual realization of their profound connection, not just to their work but to their purpose. They were integral threads in a world transforming at breakneck speed, embodying precision, purpose, and collaboration.

And yet, as the sun set, revealing a city aglow with the soft hum of progress, one truth lingered between them: Marti and Vivek were humanoid robots, the epitome of technological progress. Their journeys reflected what machines could achieve and how they could inspire a future defined by balance, efficiency, and connection.

AI, ROBOTICS, and the FUTURE OF WORK

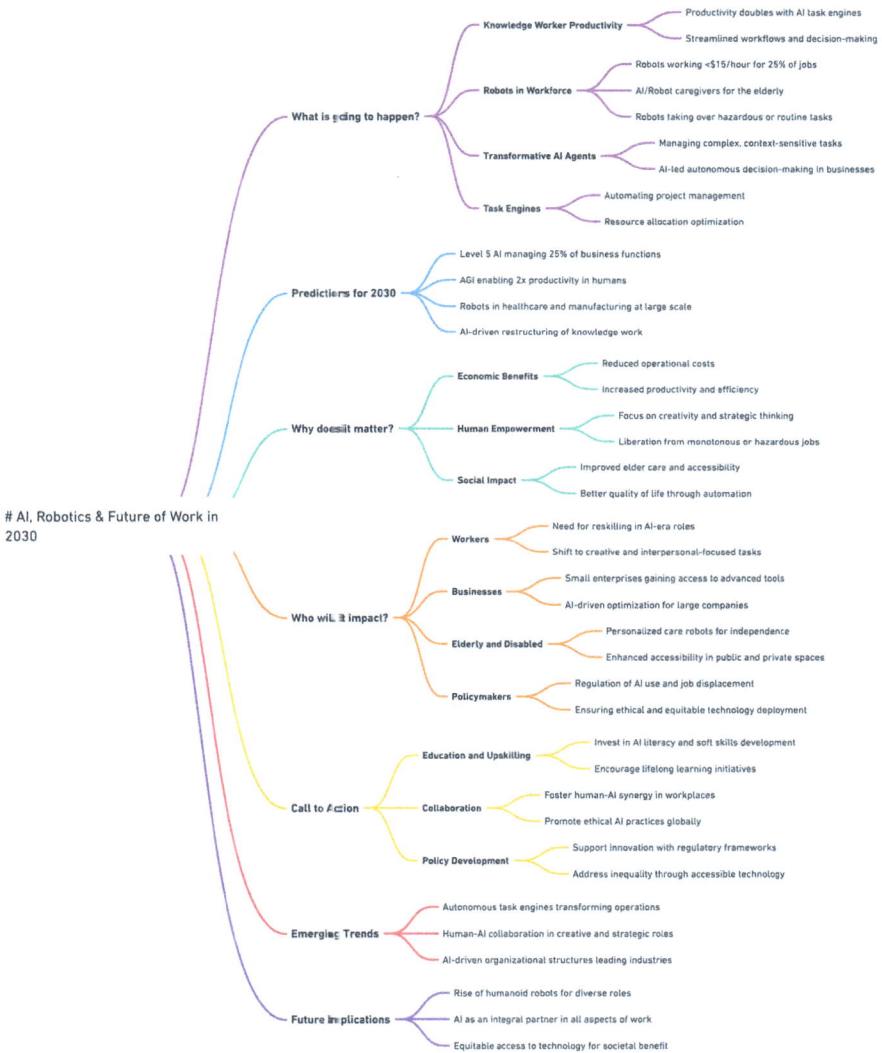

 # AI, Robotics & Future of Work in 2030

- **What is going to happen?**
  - Knowledge Worker Productivity
    - Productivity doubles with AI task engines
    - Streamlined workflows and decision-making
  - Robots in Workforce
    - Robots working <$15/hour for 25% of jobs
    - AI/Robot caregivers for the elderly
    - Robots taking over hazardous or routine tasks
  - Transformative AI Agents
    - Managing complex, context-sensitive tasks
    - AI-led autonomous decision-making in businesses
  - Task Engines
    - Automating project management
    - Resource allocation optimization
- **Predictions for 2030**
  - Level 5 AI managing 25% of business functions
  - AGI enabling 2x productivity in humans
  - Robots in healthcare and manufacturing at large scale
  - AI-driven restructuring of knowledge work
- **Why does it matter?**
  - Economic Benefits
    - Reduced operational costs
    - Increased productivity and efficiency
  - Human Empowerment
    - Focus on creativity and strategic thinking
    - Liberation from monotonous or hazardous jobs
  - Social Impact
    - Improved elder care and accessibility
    - Better quality of life through automation
- **Who will it impact?**
  - Workers
    - Need for reskilling in AI-era roles
    - Shift to creative and interpersonal-focused tasks
  - Businesses
    - Small enterprises gaining access to advanced tools
    - AI-driven optimization for large companies
  - Elderly and Disabled
    - Personalized care robots for independence
    - Enhanced accessibility in public and private spaces
  - Policymakers
    - Regulation of AI use and job displacement
    - Ensuring ethical and equitable technology deployment
- **Call to Action**
  - Education and Upskilling
    - Invest in AI literacy and soft skills development
    - Encourage lifelong learning initiatives
  - Collaboration
    - Foster human-AI synergy in workplaces
    - Promote ethical AI practices globally
  - Policy Development
    - Support innovation with regulatory frameworks
    - Address inequality through accessible technology
- **Emerging Trends**
  - Autonomous task engines transforming operations
  - Human-AI collaboration in creative and strategic roles
  - AI-driven organizational structures leading industries
- **Future Implications**
  - Rise of humanoid robots for diverse roles
  - AI as an integral partner in all aspects of work
  - Equitable access to technology for societal benefit

WHAT IS IT?

Integrating AI and robotics will redefine work by 2030, creating unprecedented shifts across industries, economies, and personal lives. Robots will handle tasks once deemed irreplaceable, from healthcare support to manufacturing operations, enabling people to focus on creativity, strategic thinking, and interpersonal interactions. These technologies won't simply replace jobs but will foster an evolution—pushing humans to leverage their unique skills while machines take over routine, dangerous, or tedious work.

This transformation represents a moment of opportunity. Knowledge workers will benefit from AI-enhanced productivity, while society will see more affordable services and improved quality of life. Innovations like task engines and AI agents will streamline operations, minimize inefficiencies, and make faster and more data-driven decisions. As roles change, workers must embrace continuous learning and develop emotional intelligence, creativity, and complex problem-solving abilities to stay relevant.

However, the shift isn't without challenges. Businesses must adopt agile strategies to integrate AI effectively, fostering human-machine collaboration that maximizes precision and human insight.

Policymakers must navigate the ethical complexities of AI deployment, ensuring responsible use and equitable access across all layers of society. Preparing for this future requires thoughtful planning, investment in education, and a willingness to embrace change.

The vision for 2030 is clear: AI and robotics will not just alter the workforce but augment it, driving innovation and efficiency while freeing people to pursue more meaningful work. As we stand on the edge of this transformation, the path forward is collaboration—between humans and machines, businesses, innovators and regulators, and individuals and lifelong learning. The future belongs to those willing to adapt, learn, and thrive in this dynamic, AI-powered world.

WHAT IS GOING TO HAPPEN?

By 2030, AI agents will likely be capable of learning and adapting to new situations without human input, mimicking cognitive functions that were once considered uniquely human. AI and robotics will fundamentally transform daily life and work. Robotic personal care assistants will support older people independently, significantly enhancing their quality of life. AI and robots, capable of operating under $15 per hour, could take many of today's jobs, shifting the workforce towards roles demanding creativity and complex decision-making. Knowledge workers will see their productivity doubled, aided by AI-driven task engines and agents that streamline workflows and decision-making processes. Task engines will automate project management and operational tasks in ways that optimize time and resources across organizations. Powered by advanced machine learning algorithms, these engines will predict project timelines, allocate resources, and even identify potential disruptions before they occur, significantly enhancing organizational agility. These technologies will redefine traditional roles and create new opportunities for innovation and efficiency in every sector of the economy.

PREDICTIONS FOR AI, ROBOTICS, AND WORK IN 2030

1. **2x Knowledge Worker Productivity Increase:** With 1 billion knowledge workers, productivity will increase significantly. This will be driven by AI agents and task engines optimizing workflows, decision-making, and time management.

2. **Robots as Cost-Effective Workers:** robots can work at less than $15 per hour, making them viable for more than 25% of current jobs. This shift will significantly alter the landscape of low-skill and routine jobs, pushing humans towards roles that require complex decision-making, creativity, and interpersonal skills.

3. **Robots for Healthcare:** the elderly can purchase an AI/Robot personal care assistant that can fully support their health and daily needs without human assistance.

4. **Transformative AI Agents:** AI agents will evolve to handle more complex and context-sensitive tasks, transforming traditional job roles. These agents will enable humans to offload routine and administrative tasks, allowing them to focus on strategic and innovative work.

5. **Proliferation of Task Engines:** Task engines powered by advanced AI will become integral in project management and operations, automating the delegation of tasks and optimizing resources. This will lead to more dynamic and responsive business environments with faster and more data-driven decision-making. This will increase productivity by >25% for >25% of industries.

DOES IT MATTER?

The transformations driven by AI and robotics will reshape the fabric of our daily lives, economies, and societies. These technologies hold the potential to revolutionize industries by enhancing efficiency, accessibility, and affordability in ways previously unimaginable. For individuals, this means liberation from monotonous, labour-intensive tasks, opening doors to more creative, strategic, and fulfilling work. People will be empowered to focus on what truly matters—be it innovation, caregiving, or personal growth—while machines handle repetitive or hazardous jobs.

Economically, AI and robotics promise a ripple effect of benefits: reduced operational costs, heightened productivity, and the creation of entirely new markets. From precision-driven manufacturing to seamless logistics, businesses can operate smarter and faster, fostering economic growth and resilience. Socially, these advancements tackle pressing challenges, such as ensuring elder care for aging populations or making public spaces more accessible for people with disabilities, significantly improving the quality of life for the most vulnerable.

However, this transformation is not just about technology—it's about humanity. It's a chance to rethink how we work, live, and care for one another.

By embracing these shifts thoughtfully, we can ensure that the immense potential of AI and robotics uplifts all corners of society, leaving no one behind in this automated era. Understanding and preparing for these changes is not just a necessity—it's an opportunity to shape a future that prioritizes connection, creativity, and compassion.

WHO WILL IT IMPACT?

The rise of AI and robotics will touch every aspect of society, reshaping lives and livelihoods across the globe. Workers in routine and manual jobs will face the most immediate impact, requiring them to reskill and pivot toward roles that demand uniquely human qualities like creativity, critical thinking, and emotional intelligence. Integrating AI will streamline tasks for knowledge workers, reducing administrative burdens and freeing them to focus on innovation and strategic thinking.

For older people and individuals with disabilities, the future holds profound promise. Personalized care robots and assistive technologies will provide greater independence, ensuring they can live more comfortably and with dignity. Families will feel relief knowing tireless, capable companions support their loved ones.

Businesses of all sizes will experience seismic shifts in operations. Small enterprises will gain access to tools once available only to large corporations, while major companies will harness AI to optimize efficiency, reduce costs, and open new avenues for growth. However, this transformation will demand agility and a willingness to adapt to changing workforce dynamics.

Policymakers will be tasked with navigating uncharted territory, from crafting regulations that ensure ethical AI use to addressing job displacement. They must ensure that technological progress benefits everyone, not just a privileged few, by fostering accessibility for all and maintaining robust safety and ethical standards.

CALL TO ACTION

To prepare for the AI-driven future, prioritize continuous learning and skill development. As AI reshapes industries and job roles, staying adaptable and technologically fluent will be crucial. Engage in educational programs focusing on AI literacy and the soft skills increasingly valued in the workforce, such as critical thinking, creativity, and interpersonal communication. Embracing lifelong learning will ensure your relevance in a rapidly changing job market and position you to leverage AI advancements to your advantage. Start today by identifying courses or workshops that align with future industry demands and commit to a path of ongoing personal and professional growth.

A DEEPER DIVE

Doug Hohulin on AI, Robotics and Work

My first job in 1983 as an engineer in training involved tackling a dull and time-consuming task. I was assigned to automate the testing of a military radio with 7,000 channels, each needing a check for self-interference (signal-to-noise level). The manual process required a technician to painstakingly tune to one channel at a time, test all seven thousand channels, and record the signal-to-noise ratio. This process—tune, test, write down results, and repeat 7,000 times—took a week to complete. Afterward, the technician would review all the data and identify which channels had self-interference issues that needed correction in the radio's design.

Instead, I developed a symbolic/knowledge-based AI program that automated this easily. The task could begin at the end of the day and be completed overnight with a report identifying which of the 7,000 channels needed fixes for self-interference. It was a small but powerful glimpse into how AI could transform laborious tasks, setting the stage for how AI and automation continue to revolutionize work today and are set to reshape it by 2030 drastically. If a job is dirty, dangerous, complex, or dull, give it to the AI or the robot.

The start on my AI work journey offered a small but profound glimpse into the future of work. In this future, monotonous tasks are relegated to machines, freeing human creativity and strategic thinking. This is a backdrop for exploring how work has evolved and will continue to transform by 2030, driven by advancements in AI, robotics, and digital technologies. Back in 1983, the workplace was predominantly manual, with computers and automation beginning to make their inroads into traditional industries.

Fast forward to 2024, and we find ourselves in an era where digital tools and AI have become central to our work environments. These technologies have not only increased productivity but also begun to reshape the very nature of work itself. Routine tasks are increasingly automated, allowing workers to focus on more complex and creative endeavours. This shift significantly affects job roles, worker skills, and organizational structures.

As we look towards 2030, the pace of change is set to accelerate with the proliferation of AI agents and task engines. These sophisticated tools will diminish the need for human involvement in routine tasks, instead empowering individuals to tackle more strategic and innovative projects. This evolution is not just about technology replacing human labour; it's about augmenting human capabilities and creating opportunities for greater creativity and strategic influence in work.

Introducing AI agents capable of handling more complex and context-sensitive tasks will transform traditional job roles across all sectors. From healthcare, where AI-assisted robots will provide care, to manufacturing, where autonomous robots will execute tasks, the work landscape will be radically different. These changes will require a new mindset and a re-evaluation of the skills valued in the workforce.

Case Study 1: What does achieving AGI in Healthcare and Aging in Place mean?

Apple co-founder Steve Wozniak once proposed that we will have achieved artificial general intelligence (AGI) when a machine can enter a stranger's home and make a pot of coffee. For me, AGI will be realized when an older adult chooses an AI/robot personal care assistant that can fully support their health and daily needs without human assistance.

Clinicians shouldn't fear job loss due to automation, AI, or robotics. With a global healthcare workforce expected to grow from 65.1 million in 2020 to 84 million by 2030, the demand for healthcare professionals will remain high, especially in underserved regions. AI and robotics will enhance, not replace, clinical roles by automating repetitive tasks, streamlining workflows, and providing decision support. These technologies help reduce burnout, allowing clinicians to focus more on patient care. Improved diagnostics, personalized treatments, and administrative relief will empower healthcare providers, ensuring better outcomes while fostering a more sustainable and fulfilling work environment for medical professionals.

The Evolution of Work

As influenced by technological advancements, the evolution of work is a narrative of perpetual transformation, characterized by a shift from manual labour to an era where AI and robotics redefine the essence of professional engagement and productivity. This progression mirrors broader socio-economic shifts and reflects our increasing reliance on technology to bolster efficiency and innovate solutions to age-old problems.

1983: The Advent of Computers in Workplaces

In 1983, the landscape of work was predominantly manual. Computers were a burgeoning technology, slowly entering mainstream workplaces; their potential was yet to be fully realized. Tasks that required extensive human effort and time were the norm, from typewriting documents to manually testing and recording data. This period marked the initial integration of technology into work, primarily aimed at automating simple, repetitive tasks that were clearly defined and structured.

2024: Rise of AI and Digital Transformation

Fast forward to 2024, the role of AI in the workplace has expanded dramatically. Digital transformation has become a strategic imperative for businesses across sectors. AI technologies have become integral to operational efficiency, from simple chatbots handling customer service inquiries to more sophisticated AI systems managing supply chains and logistics. The digital tools of this era do more than automate tasks; they will analyze vast amounts of data to inform decision-making, predict consumer behaviour, and personalize customer experiences. This shift has significantly boosted productivity by allowing human workers to focus on higher-level tasks that require creative and strategic thinking.

The Transition to 2030: Pervasive AI Integration

Looking toward 2030, the evolution of work is poised to take an even more dramatic turn. The development of AI agents and task engines suggests a future where AI does not just supplement human work but starts to replace specific roles with high levels of autonomy and efficiency. These AI systems are expected to handle complex, context-sensitive tasks that are the purview of skilled professionals today. For instance, AI in healthcare might diagnose diseases and create personalized treatment plans, while AI in finance could manage investments and prevent fraud.

Task engines will automate project management and operational tasks to optimize organization time and resources. Powered by advanced machine learning algorithms, these engines will predict project timelines, allocate resources, and identify potential disruptions before they occur, significantly enhancing organizational agility.

AI Agents and Task Engines - The New Workforce

As we edge closer to 2030, the workforce is undergoing a transformative shift, influenced heavily by the rapid evolution and integration of AI agents and task engines.

This new breed of technological tools is not merely enhancing the capabilities of human workers but is starting to take on roles traditionally reserved for skilled professionals.

The levels of AI progress according to OpenAI:

Level 1: Chatbots, AI with conversational language – we are here

Level 2: Reasoners, human-level problem solving

Level 3: Agents, systems that can take actions

Level 4: Innovators, AI that can aid in invention

Level 5: Organizations, AI that can do the work of an organization

AI Agents: Beyond Simple Automation

AI agents are advancing beyond their initial roles of providing support and handling routine tasks. These agents are now equipped with capabilities that allow them to understand and react to complex situations with nuance and efficiency that rivals human intelligence. For example, in customer service, AI agents are transitioning from answering basic queries to managing entire customer relationships, using insights drawn from vast data to personalize interactions and predict future needs.

In professional settings, AI agents are becoming indispensable by taking over tasks such as legal document analysis, where they can quickly sift through thousands of pages to extract relevant information. This process would take humans considerably longer. These agents use natural language processing and machine learning to understand the content at a granular level and provide insights and summaries that aid decision-making.

Task Engines: Optimizing Workflows

Task engines represent another pivotal element in the AI-driven transformation of the workplace. These systems are designed to automate and optimize complex workflows, ensuring that projects are completed efficiently and resources are allocated effectively. By analyzing patterns and outcomes from past projects, task engines can forecast potential bottlenecks and suggest optimal workflows, which helps minimize waste and maximize productivity.

In sectors such as manufacturing, task engines automate the scheduling and management of tasks across the production line. They dynamically adjust to changes in demand or supply chain disruptions by reallocating resources in real time, thus maintaining operational efficiency.

AI in Strategic Decision-Making

Beyond operational tasks, AI agents and task engines are increasingly involved in strategic decision-making. They synthesize information from diverse sources to provide recommendations informed by deep data analysis. For instance, in financial investments, AI systems analyze market trends, company performance data, and economic indicators to offer investment strategies with a high probability of success.

These AI-driven tools are also crucial in scenarios requiring risk assessment, where they evaluate the potential impacts of different decisions, providing executives with a comprehensive risk-benefit analysis that supports more informed and timely decisions.

Human-AI Collaboration

Integrating AI agents and task engines into the workforce does not diminish the importance of human workers; instead, it reshapes their roles. Humans are needed to oversee AI operations, provide creative insights, and make ethical decisions where AI may still lack judgment. This collaboration between human and artificial intelligence creates a new hybrid workforce where each complements the other's strengths.

Preparing for the AI-Enhanced Workforce

Organizations must prepare for this future by investing in AI literacy and infrastructure that supports seamless human-AI integration. Training programs must be revamped to focus not only on technical skills but also on enhancing soft skills such as problem-solving, critical thinking, and interpersonal communication, which are crucial for managing AI partnerships.

As we move towards 2030, AI agents and task engines are set to revolutionize the workforce, making businesses more efficient, jobs more fulfilling, and career paths more diverse. Embracing this shift will require an adaptable, forward-thinking approach to work, where continuous learning and innovation are at the forefront. This evolution promises enhanced productivity and a more profound human-machine synergy that could redefine the very essence of work.

Navigating the Middle Zone of AI and Human Collaboration

As we transition into an era dominated by advanced artificial intelligence, the "middle zone" of AI and human collaboration emerges as a critical concept as discussed by Erik Brynjolfsson at the Standford ECON295 class. This is the realm where human creativity and strategic insight synergize with AI's processing power and data-driven precision to enhance productivity, innovation, and economic growth. Understanding and optimizing this collaboration is essential for leveraging the best of both worlds—combining human beings' unique strengths with AI systems' capabilities.

In this middle zone, AI handles data-heavy tasks, processing and analyzing information at scales and speeds unattainable by humans alone. Meanwhile, humans contribute contextual understanding, ethical judgment, and creative problem-solving. This partnership is seen in industries such as healthcare, where AI algorithms assist in diagnosing diseases by analyzing medical images with superhuman accuracy.

At the same time, doctors provide a nuanced understanding of patient history and care that only human experience can offer.

To thrive in this middle zone, training programs must evolve. Workers need to develop skills that complement AI capabilities. This includes focusing on strategic decision-making, emotional intelligence, leadership, and lifelong learning. Organizations should foster environments where AI tools are seen as partners in the problem-solving process, not just as tools for efficiency. By doing so, they can cultivate a workforce that is technically proficient and adept at integrating AI insights into broader strategic goals.

Navigating the middle zone effectively means embracing AI as a part of the team, where each member—human or machine—plays to their strengths. This balanced approach promises to maximize the benefits of AI while maintaining the irreplaceable value of human insight. As we look toward 2030, the success of businesses and the health of economies will increasingly depend on how well we can manage this powerful collaboration between humans and artificial intelligence.

Preparing for the Future of Work

The future of work will demand higher adaptability and a strategic embrace of technological advancements. Here's how various stakeholders can prepare to thrive in this evolving landscape.

1. **Embracing AI Integration**

    Organizations must start by fully integrating AI into their business processes. This goes beyond merely adopting new tools; it involves a cultural shift towards innovation and continuous improvement. Leaders should foster an environment where experimenting with AI solutions is encouraged, and failures are seen as learning opportunities. This proactive approach will help companies stay competitive and lead in adopting emerging technologies that can transform their operations.

2. **Upskilling and Reskilling**

For individuals, continuous learning is the key to remaining relevant in the AI-enhanced workplace. As AI takes over routine tasks, the human work-force needs to upskill or reskill to pivot into roles that require complex problem-solving, emotional intelligence, and creative thinking—skills that AI cannot easily replicate. Educational institutions and employers should provide training programs focused on these areas and how to effectively collaborate with AI systems.

3. **Fostering Human Collaboration**

Organizations should design workflows that facilitate effective collaboration between humans and AI. This includes creating roles that bridge the gap between technical AI management and business operations, ensuring that AI tools enhance decision-making rather than replace human judgment. Companies should invest in interface design that makes AI tools accessible and understandable to all employees, regardless of their technical expertise.

4. **Ethical and Responsible AI Use**

It's crucial to address the ethical implications of AI in the workplace. Organizations must establish clear policies and guidelines to govern AI use, ensuring that these technologies are used responsibly and do not exacerbate biases or lead to unequal treatment of employees. This also includes compliance with evolving regulations around data privacy and AI transparency.

5. **Strategic Planning for AI Impact**

Finally, strategic planning for AI's impact on business models and employment patterns is essential. Leaders should consider how AI could disrupt their industry and proactively plan for market dynamics and skill demand changes. This might involve diversifying service offerings or developing new business segments that leverage AI capabilities to meet emerging needs.

By taking these steps, individuals and organizations can position themselves to survive and thrive in the rapidly approaching future of work shaped by AI and robotics. The vision for 2030 is clear: AI and robotics will not just alter the workforce but augment it, driving innovation and efficiency while freeing people to pursue more meaningful work. As we stand on the edge of this transformation, the path forward is collaboration—between humans and machines, businesses, innovators and regulators, and individuals and lifelong learning. The future belongs to those willing to adapt, learn, and thrive in this dynamic, AI-powered world.

Metaverse, XR, Spatial Computing, Computer Vision, and BCI

Julie and Holly maintained their close bond despite living drastically different lives. An extended reality (XR) pioneer, Julie spent her days building platforms and tools that revolutionized industries such as training, marketing, retail, and entertainment. Meanwhile, Holly, her adventurous daughter, travelled the world, leveraging XR glasses and AI-powered technologies to keep learning and staying connected to her mother.

The day began with a shared moment: working out together. Julie, in her Toronto home, put on her XR glasses and joined Holly, virtually, who was overlooking the serene beaches of Bali. Their glasses created a shared holographic workout studio where their AI trainer guided them through yoga and strength exercises. The environment, enriched with spatial computing, mimicked a peaceful mountaintop with dynamic visuals that adapted to their pace. Holly laughed when her AI suggested a stricter workout for Julie, who said, "Maybe I'll stick to meditation next time."

After their session, Holly headed off to explore local markets, her glasses acting as her real-time guide and mentor. Cameras and microphones embedded in her glasses captured the sights and sounds around her. At the same time, AI analyzed everything she saw, offering cultural insights, translation help, and even nutritional facts about the exotic fruits she picked up. The glasses also logged her learning journey, creating a personalized, blockchain-secured record of her experiences.

Julie, on the other hand, immersed herself in her work. As the CEO of an XR platform company, she spent her morning refining a new virtual world designed for corporate training. Her XR workspace transformed her physical office into a collaborative hub, where her team from around the globe joined as avatars to finalize a client presentation. Julie's tools allowed her to conjure 3D models, test immersive scenarios, and integrate gamified elements into serious applications—all in real time.

Despite the distance, Julie and Holly stayed connected. During Holly's lunch break, she hosted a virtual gathering in her personal metaverse space—a vibrant fusion of a coffee shop and a futuristic lounge. Julie joined as an avatar, catching up with Holly and her global network of friends. Holly's space allowed for seamless collaboration and socializing, where everyone's avatars reflected their unique personalities.

Later in the day, Julie received a message from Holly's AI assistant. Holly had stumbled upon a unique architectural design in Bali and thought Julie might find inspiration for her next XR project. The glasses' AI captured and rendered the design into a 3D model, which Julie reviewed and incorporated into a virtual retail experience she was developing.

As the sun set, both Julie and Holly reflected on their interconnected lives. Holly marvelled at how her glasses turned every moment into a learning opportunity, helping her navigate a rapidly changing world. Julie felt pride in knowing the tools she developed enabled meaningful experiences like the ones she shared with Holly. Despite being miles apart, technology allowed them to stay close, learn from one another, and thrive in their respective journeys.

What neither of them could have imagined years ago was how seamlessly technology would bridge physical distances and the gaps between their very different worlds. For Julie and Holly, the future wasn't just about innovation but about connection, growth, and shared experiences. Together, they exemplified how technology could unite, inspire, and empower humanity one day at a time.

XR IN 2030

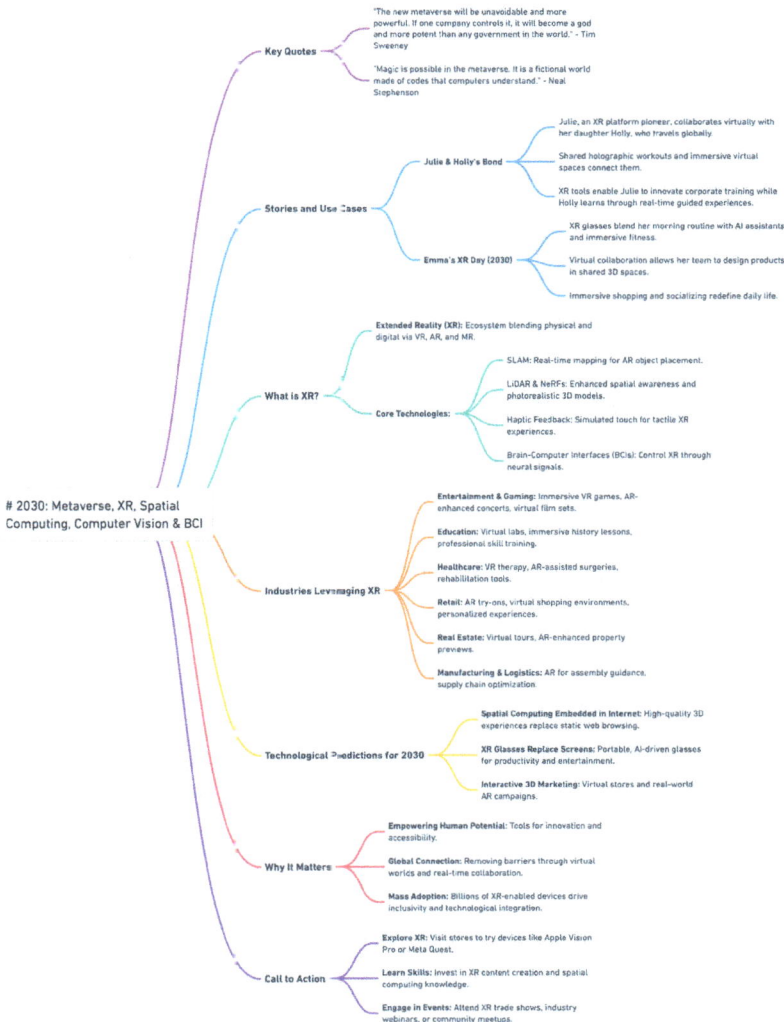

ALAN SMITHSON ON XR

The world is on the cusp of a revolutionary transformation. The convergence of cutting-edge technologies is giving rise to a new era of human-computer interaction that promises to redefine the boundaries of human experience. In this chapter, we will explore the exciting landscape of the metaverse, virtual, augmented and mixed reality (XR), spatial computing, computer vision, and brain-computer interfaces (BCI). These technologies are poised to disrupt industries, reshape how we interact with the world, and unlock new possibilities for human innovation.

I have been fortunate to have been part of the birth of the XR industry, having started working in the field in 2014 after trying the DK1 at Curiosity Camp hosted by former CEO of Google, Eric Schmidt. Since that fateful day, I realized the genuinely transformative potential of head-worn devices, now called spatial computing. Our team has produced over 200 XR projects for some of the world's largest companies. I hosted the XR For Business Podcast to understand the use cases, challenges and opportunities this new technology would unlock.

In April 2016, we renamed our company after an obscure reference to virtual worlds in a book published in 1992 called 'Snow Crash.' We named our company METAVRSE (minus the second E); the rest is history. The Metaverse has many definitions, but it refers to a future state of the internet that leverages 3D, XR, AI and Blockchain. Our product, METAVRSE, is a 3D creation platform on the web that makes it easy for anyone to build applications and virtual worlds across myriad enterprise and consumer applications such as entertainment, gaming, training, retail, marketing, and collaboration. My wife and business partner, Julie Smithson, has been the President of the VR/AR Association (Toronto Chapter) and is the Founder of XR Women, a global community of women leaders in spatial computing and she was named one of the Top 100 Women of the Future in 2024. Julie and I are honoured to be mentored by the Grandfather of VR, Thomas Furness, who pioneered the cockpit heads-up display for the US Air Force in the 1960s. We are humbled to be pioneers of the

XR industry from the beginning; our knowledge and experience have informed this chapter of the book.

By 2030, nearly everyone will have access to an internet-connected device, enabling them to contribute valuable work from anywhere. Virtual meeting spaces, whether experienced in immersive headsets or through familiar 2D interfaces, will allow people to represent themselves as avatars, transcending geography, gender, race, and identity. Avatars will become extensions of who we are—or who we want to be—adapting seamlessly from business meetings to gaming sessions with friends. With just a few clicks, we will craft our ideal representation for any situation, empowering individuality and creativity like never before.

The ability to live, work, and collaborate in virtual worlds will break down barriers of language and borders, fostering a truly global society. Combined with AI-driven tools, spatial computing will render the world's data into visually immersive and interactive environments, transforming how we learn, work, and connect. The metaverse—a natural evolution of the internet—will allow us to build and traverse entirely new dimensions, own and transfer digital assets, and instantly create custom virtual worlds with AI. Imagine texting a simple prompt and conjuring a virtual training environment or uploading a manual to generate a spatial learning experience in real-time. This is no longer science fiction; companies like Siemens and Bell Helicopter are already proving how these technologies are accelerating innovation at unprecedented speeds.

By 2030, breakthroughs like multifocal depth planes will make virtual environments feel as accurate as the physical world, while brain-computer interfaces will connect millions of people directly to the internet. These advancements will help the paralyzed communicate and move, revolutionize how we operate machines, and unlock unimaginable potential in human creativity and collaboration.

The ultimate goal is to empower every human on Earth with the tools to achieve unlimited potential, transcend limitations and work together as one global society. This is not just the future of technology—it's the future of humanity, united to build a world where everyone has the superpowers to create, innovate, and thrive together.

WHAT IS IT?

Extended reality (XR) is an ecosystem of technologies that blur the line between physical and digital experiences. XR includes virtual reality (VR), where users immerse fully in computer-generated worlds; augmented reality (AR), where digital overlays enhance the real world; and mixed reality (MR), where virtual and physical elements interact seamlessly. These technologies, called immersive or spatial computing, enable intuitive, human-like interactions with computers, fundamentally reshaping how we work, play, and connect.

The Technology of XR

XR relies on a convergence of groundbreaking technologies, each contributing to the depth, realism, and functionality of these immersive environments:

Simultaneous Localization and Mapping (SLAM): XR devices can understand and map their surroundings in real time, anchoring digital objects in physical spaces. This technology is essential for AR applications, such as projecting a virtual guide on your coffee table or accurately placing 3D objects in real-world environments.

Computer Vision: This technology empowers XR devices to interpret visual data from the real world. It recognizes objects, gestures, and faces, enabling natural and responsive interactions between users and digital elements.

Photogrammetry: By analyzing photos taken from various angles, photogrammetry reconstructs realistic 3D models of real-world objects or locations. This technology lets users explore digital twins of monuments, cities, or interiors in great detail.

Light Detection and Ranging (LiDAR): This technology uses light pulses to measure distances and create precise 3D maps. It enhances spatial awareness, and is built into devices such as the iPhone and high-end XR headsets, making virtual objects interact convincingly with the physical world.

Neural Radiance Fields (NeRFs) and Gaussian Splats: NeRFs use AI to create realistic 3D models with intricate lighting and shading from just a few images. Gaussian splats, on the other hand, are an efficient way to render visuals, ensuring high-quality graphics with minimal computational load.

Haptic Feedback: This technology simulates touch and physical sensations, creating a tactile dimension in XR. Whether feeling the texture of a virtual object or experiencing the recoil of a digital tool, haptics deepens immersion. Locomotion systems in XR allow users to navigate virtual spaces naturally, whether through physical movement, simulated environments, or advanced techniques such as treadmills or teleportation mechanics.

The People and Characters of XR

Avatars: Avatars serve as digital representations of users, customizable to reflect personal identity, style, or mood. They enable immersive interactions in social platforms, professional meetings, and collaborative projects.

Non-Player Characters (NPCs): NPCs are AI-driven entities that populate XR worlds, acting as guides, collaborators, or interactive companions. With advances in natural language processing (NLP), NPCs engage users in dynamic conversations and respond contextually to commands or queries.

Natural Language Processing (NLP): NLP bridges human and machine communication in XR, allowing users to interact with environments, avatars, and NPCs using natural speech. This transforms voice commands into seamless experiences.

The XR Continuum

- VR: Fully immersive digital worlds that isolate users from the real environment.

- AR: Enhancing the real world with contextual digital overlays.

- MR: Merging digital and real-world elements that respond to each other dynamically.

- Brain-Computer Interface (BCI): Directly connecting the human brain to a computer.

Imagine slipping on lightweight AR glasses that combine SLAM, computer vision, and LiDAR. Instantly, your room transforms. Holographic screens appear before you, your AI-powered avatar offers a morning briefing, and NeRF-rendered visuals bring depth to your workspace. Haptic gloves let you feel the texture of virtual tools, while locomotion systems transport you effortlessly across simulated environments. When you need help, an NPC assistant responds with realistic voice-based guidance powered by NLP. A BCI translates your thoughts into actions for tasks requiring ultimate precision, redefining how you control the digital realm.

From recreating landmarks and preserving historically significant places and things with photogrammetry to collaborating in immersive virtual offices, XR merges technology into life's fabric, making the once unimaginable a daily reality. Welcome to the future of human interaction.

Brain-Computer Interfaces (BCIs): BCIs enable direct communication between the human brain and digital systems. By detecting and interpreting neural signals, BCIs allow users to interact with XR environments through thought alone, promising a future of hands-free control and unparalleled accessibility. BCIs enable humans to interact with computers using their brain signals. BCI can potentially revolutionize how we interact with technology, particularly for individuals with disabilities. There are two types of BCI, each with its unique characteristics and applications:

1. **Invasive BCI:** Requires surgical implantation of electrodes in the brain.

2. **Non-invasive BCI:** Uses sensors and algorithms to detect brain signals without surgical intervention.

By 2030, BCI technology will have advanced significantly, leading to transformative use cases across various fields:

1. **Medical Rehabilitation:** BCIs will aid in rehabilitating stroke patients by enabling them to control robotic limbs or exoskeletons with their thoughts, accelerating recovery and improving mobility.

2. **Assistive Technology:** Non-invasive BCIs will allow individuals with severe physical disabilities to communicate and control their environments more effectively, such as operating smart home devices or communicating through speech-generating devices.

3. **Mental Health:** BCIs will be used to monitor and modulate brain activity for mental health treatments, providing real-time feedback and interventions for conditions such as depression, anxiety, and PTSD.

4. **Enhanced Learning:** BCIs will facilitate personalized learning experiences by adapting educational content based on the user's brain activity, optimizing cognitive load, and improving retention.

5. **Entertainment and Gaming:** BCIs will revolutionize the gaming industry by enabling players to control games with their minds, creating immersive experiences where thought and intention drive gameplay dynamics.

Industries Leveraging XR

- **Entertainment and Gaming:** In gaming, VR titles such as Half-Life: Alyx transport players to richly detailed virtual worlds, while AR games such as Pokémon Go bring gameplay to real-world settings. Virtual concerts, hosted on platforms such as Wave, allow fans to attend performances in fantastical environments without leaving home. Filmmakers leverage XR to craft previsualizations or even entirely virtual films, seamlessly blending real and digital elements.

- **Education:** Virtual classrooms allow students to explore ancient ruins, dive into underwater ecosystems, or traverse the human body for anatomy lessons, all without leaving their school. Platforms such as zSpace and ENGAGE XR provide hands-on virtual labs and immersive simulations for STEM subjects. Professional training benefits from XR, too, with realistic simulations for fields such as medicine, aviation, and engineering, where learners can practice complex skills in safe, controlled environments.

- **Healthcare:** Virtual surgery simulations allow surgeons to practice procedures in risk-free environments, improving precision and outcomes. AR assists surgeons during operations, directly overlaying critical data and 3D models onto the patient's anatomy. For mental health, VR therapy offers effective treatments for anxiety, PTSD, and phobias by immersing patients in controlled virtual environments. Rehabilitation programs utilize XR to help patients regain mobility and function through engaging, gamified exercises.

- **Retail and E-commerce:** Virtual try-ons let retail and e-commerce customers see how clothes, accessories, or makeup look on them without stepping into a store. At the same time, AR apps enable shoppers to visualize furniture in their homes before purchasing. Virtual stores provide immersive browsing experiences, letting customers explore digital showrooms and interact with products in 3D.

- **Architecture and Construction:** Digital twins allow architecture and construction professionals to create and explore virtual replicas of buildings or entire cities, identifying potential issues before construction begins. Virtual walkthroughs let clients experience designs in 3D, enabling feedback and changes in real time. On-site workers use AR to overlay blueprints onto physical environments, streamlining tasks and reducing errors, while construction managers leverage XR to train workers in safety procedures and equipment use.

- **Manufacturing and Logistics:** On manufacturing assembly lines, AR guides workers with step-by-step instructions displayed directly on components, reducing errors and training time. Maintenance teams use AR to visualize machine parts and receive real-time diagnostics, enabling faster repairs. In warehouses, AR aids in inventory management by highlighting the locations of items and suggesting efficient pick paths. Logistics companies utilize XR simulations to optimize supply chains and enhance delivery teams' training.

- **Military and Defense:** VR combat simulations immerse soldiers in realistic scenarios, helping them practice tactics and decision-making in safe environments. AR overlays enhance battlefield awareness, displaying maps, enemy positions, and mission-critical information directly in a soldier's field of vision. XR is also used to train pilots and operators of complex machinery through lifelike simulators, ensuring readiness without the risks of real-world exercises.

- **Real Estate:** XR makes buying and selling real estate more efficient and engaging. Virtual property tours allow potential buyers to explore homes remotely, saving time and expanding access to international markets. AR enhances in-person viewings by letting clients visualize renovations or see how spaces would look with different furniture. Developers use XR to showcase properties still under construction, providing lifelike previews that help secure investments and sales before projects are complete.

- **Marketing and Advertising:** Brands use AR to enable interactive product experiences, like trying on virtual sunglasses or seeing how a car would look in their driveway. Virtual showrooms and branded metaverse spaces allow companies to showcase products innovatively, fostering deeper customer connections. Through XR advertisements or experiences, immersive storytelling builds emotional resonance with consumers, turning marketing into an engaging, participatory journey.

3D and XR Creation Platforms and Game Engines

By 2030, 3D and XR creation platforms and game engines will be at the heart of content development across industries, powering immersive experiences, virtual worlds, and interactive applications. These tools will allow creators, developers, and businesses to design, build, and deploy rich, spatial content that seamlessly blends digital and physical environments.

3D and XR creation platforms are software ecosystems that enable users to develop virtual environments, interactive 3D models, and immersive experiences for AR, VR, and MR. Game engines are integral to this landscape, offering advanced rendering, physics, and AI capabilities to create lifelike simulations and interactive scenarios.

These platforms are democratizing the creation of immersive content. Once limited to large studios and specialized developers, the tools are now accessible to independent creators, educators, and businesses. With user-friendly interfaces, no-code options, and powerful AI-driven design features, creating in 3D or XR no longer requires extensive technical expertise.

Crafting 3D and XR content will become deeply intuitive, with AI seamlessly integrated into every stage. From initial concept sketches to the final rendering, creators will interact with platforms anticipating their needs and offering real-time suggestions, automation, and enhancements.

- **Asset Creation:** AI-assisted design will augment traditional modelling workflows, enabling creators to generate lifelike 3D assets in seconds. Whether designing a character, environment, or product, AI tools will translate simple prompts into fully realized models with textures and animations.

- **World Building:** Game engines will allow for the procedural generation of complex environments, with creators setting parameters while AI handles intricate details such as terrain, lighting, and physics. Entire ecosystems will be built with just a few clicks, leaving more time for fine-tuning and storytelling.

- **Animation and Interaction:** Animating characters and objects will no longer require frame-by-frame adjustments. AI motion-capture tools will allow creators to input basic movements or intentions, automatically generating fluid, realistic animations.

- **AI Integration:** AI-powered tools will automate complex tasks such as asset generation, scene design, and animation, reducing development time and costs. Text-to-3D tools will let creators describe a scene or object in natural language, with the platform generating it in real time.

- **Real-Time Collaboration:** Cloud-based XR platforms will allow global teams to collaborate in shared virtual environments, enabling faster content creation and iteration.

- **Photorealism and Physics:** Advances in rendering and simulation will make virtual worlds indistinguishable from reality, enhancing immersion and interactivity.

- **Adaptive Learning:** Platforms will learn individual creators' styles, offering personalized shortcuts and suggestions based on past projects.

While technology will simplify technical challenges, human imagination will remain at the heart of creation. The role of 3D and XR platforms will amplify that imagination, giving creators the tools to push boundaries and redefine what's possible in digital storytelling and design. By 2030, these platforms won't just be tools—they'll be partners in creativity, empowering a new generation of artists, developers, and visionaries to shape the immersive worlds of the future.

WHAT IS GOING TO HAPPEN?

By 2030, head-worn devices such as AR glasses and VR headsets will dominate technical training and remote collaboration, transforming industries and workflows. These devices will deliver interactive, immersive learning experiences beyond static videos or manuals, offering hands-on virtual simulations for equipment maintenance, surgery, or engineering design tasks. The productivity gains from these tools will be undeniable, driving widespread adoption across sectors.

Remote collaboration will evolve into spatially immersive environments where teams can meet, interact, and work as if they were in the same room. Holographic representations and real-time language translation will make global teamwork more seamless and personal, removing geographical and linguistic barriers. These virtual spaces will become as ubiquitous as today's video conferencing tools but far more engaging and efficient.

By the end of the decade, interactive 3D marketing and websites will revolutionize how we engage with brands and digital content. Instead of scrolling through static web pages, users will step into virtual worlds crafted by companies to showcase their products, services, and stories. Imagine walking through a virtual car showroom, customizing a vehicle in real-time, or exploring a travel destination digitally before booking. These experiences won't just stay confined to the screen—AR will allow these virtual elements to blend into the real world, creating pop-up experiences in homes, stores, and public spaces.

This shift will redefine how we learn, work, and interact, making technology not just a tool but an immersive part of our everyday lives. The convergence of AR, VR, and interactive 3D content will push the boundaries of creativity, collaboration, and commerce, opening doors to a limitless and interconnected future.

WHY DOES IT MATTER?

The investments by major companies such as Meta, METAVRSE, Google, Apple, Samsung, and NVIDIA—amounting to hundreds of billions of dollars—underscore the profound potential of spatial computing to reshape our world. These technologies are not just upgrades; they represent a paradigm shift in how humans learn, communicate, and interact with digital content.

Spatial computing will become the next primary computing medium, following the transitions from desktop to mobile. Its ability to seamlessly integrate the digital and physical worlds will impact nearly every facet of human life. Education will be transformed as interactive 3D environments make learning more immersive and engaging. Healthcare will benefit from spatial simulations for medical training and remote surgeries. Communication will evolve as meetings shift from flat screens to shared virtual spaces, enhancing collaboration and emotional connection.

This technology will also redefine retail, entertainment, manufacturing, and architecture industries, enabling unparalleled personalization and interaction. The shift to spatial computing isn't just about convenience—it's about unlocking human potential by creating intuitive, immersive, and fundamentally transformative tools.

The massive investment and rapid development signal a future where spatial technology is as ubiquitous and essential as smartphones today, making it imperative for businesses, governments, and individuals to adapt and embrace this new era.

WHO WILL IT IMPACT?

The shift to 3D and XR technologies will touch nearly every human life, transforming how we communicate, collaborate, and connect with the world around us. As of this writing, billions of devices are already 3D and XR-enabled, and platforms such as Snapchat have demonstrated the mass appeal of AR with more than one trillion AR "snaps" shared globally. This is just the beginning. Have you tried bunny ear filters on Snap, Insta, Zoom, or Meets?

By 2030, spatial computing and 3D interfaces will become as ubiquitous as smartphones today, embedding themselves into every aspect of daily life. Whether through lightweight AR glasses, XR-enabled smartphones, or immersive headsets, every person on Earth will often interact with XR technology without realizing it. From children learning in virtual classrooms to professionals collaborating in digital workspaces to families sharing experiences across continents, XR will be the medium that brings people closer together.

Communication will become richer and more dynamic. Imagine sharing not just a photo or video but a fully immersive moment—a 3D hologram of a shared experience or a virtual model of a design that others can manipulate in real time. Collaboration will transcend borders, with global teams working side by side in shared virtual environments, solving problems and creating innovations together.

This transition will enhance how we interact and make these tools inclusive and accessible. XR will become a bridge for people with disabilities, offering new ways to see, hear, and engage with their surroundings. It will democratize design and creativity, empowering individuals to build and share 3D content quickly.

As the world moves to spatial computing, everyone will become a participant in the XR ecosystem—whether as a creator, user, or innovator. The tools that once seemed futuristic will feel as natural as sending a text or making a call. In this new era, the impact will be profound and universal, redefining how humanity interacts with technology and one another.

CALL TO ACTION

Visit a store and experience XR devices such as the Apple Vision Pro or Meta Quest. Ask to try productivity apps so you can fully understand the profound implications of these technologies. Explore immersive productivity apps or attend a virtual event. Check out free events by organizations such as the VR/ AR Association and XR Women. You can also participate in industry trade shows such as Augmented World Expo and VR Days. Think beyond entertainment—test tools that enhance work or health. Invest in skills and learn about XR content creation, AI, and spatial computing to prepare for future opportunities. Apple Vision Pro demos are available in most stores in North America.

PREDICTIONS FOR 2030

- **Spatial computing will be built into the internet:** Similar to 3D video games, websites will leverage high-quality 3D rendering, allowing companies to create virtual stores, worlds, games, and apps directly into their website. Imagine shopping for Nike in VR with virtual AI Shaq and Jordan giving you their opinions on your new sneakers, pushing all your dopamine buttons simultaneously. Spatial computing will become as common as smartphones, embedded in every aspect of daily life. Advanced AR glasses or contact lenses will provide continuous, real-time digital information overlays onto the physical world, transforming how we interact with our environment. This will drive innovation in navigation, communication, retail, and even personal health, with AI-powered virtual assistants guiding our every move.

- **Wearable XR glasses will start to replace screens:** Glasses and goggles will be light and powerful enough to replace desktop screens for extended productivity, collaboration, and entertainment. Glasses will have an AI assistant agent available to you 24/7, the glasses will have cameras watching the world around you, and the AI will have the context of where, who, and what you are doing and be able to coach you in real time.

- **Remote work & education will be more immersive:** Covid proved work from home works, but we also crave meaningful encounters with our fellow workers for a human connection. As virtual avatars and interactions improve, the value they bring in terms of immersive remote work and education cannot be overstated. With a $300 pair of glasses, you will be able to work anywhere in the world, with access to the best resources.)

PREDICTIONS FOR 2040

- **Haptic gloves and skin:** Ultra-thin haptic gloves and suits will be deployed for training, where safety and workplace hazard training saves lives. Being able to reach out and pick something up (and feel it) in a virtual world and having a robot on the other side of the globe mimic your actions opens up vast opportunities.

- **The realization of the metaverse:** Brands and individuals will be able to own a piece of the digital world, and AI will make building a complex game as easy as conjuring the environment you need.

- **Redefinition of human interaction:** Shared XR experiences that convey visuals, audio, and emotions, and sensations will transcend traditional communication. Social gatherings, meetings, and events will occur in hyper-realistic virtual spaces, with avatars representing people in ways that reflect their personalities and emotions. This will lead to new forms of social dynamics, empathy, and connection, fundamentally changing how people relate to each other across distances.

Space

Enzo and Karen. Intertwined by the cosmic frontier, their lives were dedicated to pushing the boundaries of what was possible.

Enzo, a brilliant young rocket scientist based in Paris, France, worked tirelessly for a private aerospace company specializing in reusable heavy rockets. These rockets were the backbone of humanity's space aspirations, designed to carry satellites, cargo, and even humans into orbit and beyond. Enzo's days started early, and involved virtual and physical collaboration with his global team. He could visualize complex rocket components using augmented reality glasses, overlaying simulations on physical prototypes. "Every millimetre counts," he often said, as he fine-tuned designs with the help of AI systems that could predict stress points and optimize materials.

His passion, however, was fueled by something more profound. "We're not just building rockets," he would tell his colleagues. "We're building the bridges to the next chapter of human existence." For Enzo, the work wasn't just about launching payloads into space—it was about enabling a future where humanity could live and thrive beyond Earth.

Halfway across the globe in Sharjah, UAE, Karen was immersed in her groundbreaking work. An astrophysicist at the Sharjah Research Park, she was part of an international initiative mapping reachable asteroids for mining. Her mission was to identify celestial bodies rich in rare minerals that could revolutionize industries on Earth and fuel further space exploration. Karen's team could pinpoint asteroid compositions with astonishing precision using next-generation telescopes and AI-driven data analysis. "We're finding gold mines in the sky," Karen often remarked, though she quickly emphasized the importance of sustainability. "This isn't about exploitation—it's about creating a balance between our needs and the cosmos."

Karen's days blended rigorous research and collaboration with global partners. Her work often overlapped with Enzo's; many of the asteroids she mapped were targets for missions enabled by the rockets Enzo helped design. Though they had never met in person, their paths frequently crossed in virtual conferences and shared project dashboards. Karen admired Enzo's relentless drive, while Enzo often cited Karen's meticulous research as the foundation for many of his team's mission parameters.

One morning, as Enzo calibrated a new rocket system slated for an asteroid mining mission, he received a message from Karen. "The asteroid K-792," she wrote, "holds the rare isotopes we need for next-gen ion propulsion. If your team can launch within the next 18 months, we can secure a stable orbit for extraction." Enzo smiled. The timeline was tight, but the challenge ignited his competitive spirit.

Their collaboration deepened over the months. Enzo's rockets began incorporating advanced materials sourced from asteroid mining, while Karen's research benefited from the data collected by Enzo's missions. Together, they were shaping a future where space was not just a destination but an integral part of human life.

In their quieter moments, both reflected on how far humanity had come. Enzo thought about his childhood, watching grainy footage of rocket launches and dreaming of being part of something bigger. Karen remembered gazing at the stars from her family's desert home, wondering what lay beyond. Now, they were living those dreams, not as solitary explorers but as part of a vast, interconnected effort to ensure humanity's survival and prosperity.

Their work was a testament to the power of collaboration, innovation, and the unyielding human spirit. Together, Enzo and Karen were not just mapping the future—they were building it, one rocket, one asteroid, and one shared vision at a time.

SPACE IN 2030

WHAT IS IT?

As we embark on a new decade, the space industry is poised for a transformative shift. Gone are the days of government-dominated space exploration, where a handful of nations controlled the narrative and resources. The future of space exploration is now a thriving ecosystem of commercial and collaborative ventures driven by innovation, entrepreneurship, and a shared vision for the cosmos.

By 2030, the global space economy is expected to experience exponential growth, potentially reaching $2 trillion or more, fueled by advancements in key areas such as satellite technology, space tourism, lunar exploration, asteroid mining, and space-based solar power.

Space exploration in 2030 has undergone a profound transformation, shifting from a domain once dominated by government space agencies such as NASA and ESA to a thriving ecosystem of commercial and collaborative ventures. It's no longer just about planting flags or conducting scientific experiments—it's about creating practical, scalable solutions for humanity's future. Space exploration now encompasses diverse activities, from launching satellites that power our global communications and internet infrastructure to planning human settlements on the Moon and beyond.

OPPORTUNITIES FOR SPACE IN 2030:

1. **Launch Services:** Spaceports will provide access to space for satellites, spacecraft, and other payloads.

2. **Satellite Manufacturing:** Satellites will need to be manufactured for various applications, including telecommunications, Earth observation, and navigation. We will require a constellation of Earth observation satellites to provide critical data for climate monitoring, natural disaster response, and environmental management.

3. **Space Tourism:** Suborbital and orbital space tourism experiences will offer a unique opportunity for individuals to experience space travel.

4. **In-Orbit Servicing:** We will service and maintain satellites in orbit, extending their lifespan and reducing the need for new launches.

Advances in SmallSat Technology

The rise of SmallSats, compact and low-cost satellites, is revolutionizing access to space. These small, agile satellites can be built and launched quickly, making them ideal for applications ranging from Earth observation and communication to scientific research and technology testing.

SmallSat constellations enable a wide array of space-based services, including precision agriculture, disaster response, and global connectivity. Their affordability and accessibility have opened space to startups, research institutions, and even developing nations, fostering innovation and entrepreneurship on an unprecedented scale.

SmallSats' agility also allows for rapid development and deployment of new technologies, driving a new era of space innovation. Their ability to deliver cost-effective solutions reshapes industries and democratizes the space economy, bringing opportunities to more people and organizations worldwide.

Space Tourism Takes Off

Private companies have made space more accessible, driving innovation in reusable rockets, drastically reducing costs, and next-generation propulsion systems, paving the way for interstellar exploration. Space tourism is no longer a distant dream—ordinary people are taking trips to low Earth orbit, experiencing the awe of viewing Earth from space.

As humanity reaches new heights in space exploration, space tourism has emerged as an exciting and transformative frontier. Once the exclusive domain of astronauts and scientists, space travel is now within reach for everyday people, thanks to technological advancements and the rise of private space companies. Imagine experiencing the thrill of crossing the edge of our atmosphere, looking down at Earth from low Earth orbit—a dream rapidly becoming reality.

The journey of space tourism began in 2001 when a group of paying passengers boarded a Russian Soyuz spacecraft and reached an altitude of 100 kilometres. Since that historic flight, private companies such as SpaceX, Blue Origin, and Virgin Galactic have entered the scene, offering suborbital and orbital flights that promise to redefine what travel means. Today, we're witnessing the dawn of a new era where space isn't just a destination for scientists but an adventure for anyone with the curiosity and courage to explore.

A few key factors drive the rapid growth of space tourism. Advancements in reusable rocket technology and innovations in spacecraft design have significantly lowered the cost of space travel, making it more accessible. Meanwhile, space-themed entertainment—think blockbuster movies and binge-worthy TV shows—has sparked public fascination with the cosmos. But perhaps the biggest draw is the sheer wonder of it all: the chance to see Earth as a fragile, stunningly beautiful blue orb suspended in the vastness of space is a once-in-a-lifetime experience that captivates the imagination of millions.

One of the most striking aspects of space tourism is the view of the Earth from space. The planet's curvature, blue oceans, and white clouds create a breathtaking panorama unlike anything experienced on Earth. The sense of perspective and scale is profound, and it's not uncommon for passengers to feel a deep connection to the planet and its inhabitants.

Space tourism isn't just about thrill-seekers; it's a sign of how far we've come and are willing to go. This new frontier has the potential to inspire, innovate, and connect us to something far more significant than ourselves—a perspective that only space can offer.

Permanent Lunar (Moon) Base

The Moon, with its craters, mountains, and stark, barren beauty, has always captured human imagination. But in the 21st century, the Moon is no longer just a poetic muse or a scientific curiosity—it's a strategic asset for humanity's future. A permanent moon base would serve as a stepping stone for deeper space exploration and as a hub for developing cutting-edge technologies, conducting scientific research, and even safeguarding humanity from global catastrophes.

Establishing a permanent presence on the Moon is no small feat. The lunar environment is harsh and unforgiving, with no breathable air, extreme temperature swings, and relentless radiation. Overcoming these challenges will demand unprecedented levels of innovation and international collaboration. However, the rewards are immense. Tackling these obstacles will lead to breakthroughs in habitat engineering, life support systems, and sustainable energy generation—advancements that will benefit not just space exploration but also life on Earth.

Habitat design will be at the heart of the moon base's success, ensuring the safety and well-being of its residents. These habitats will feature advanced radiation shielding, temperature regulation, and systems to recycle air, water, and waste. Autonomous 3D printing technologies will use lunar regolith to construct durable, sustainable structures. Inhabitants will live in bright, modular environments equipped to maintain a stable atmosphere and grow food, creating a self-sustaining ecosystem.

Energy generation and storage will be another cornerstone of the moon base. Solar panels, strategically placed to maximize exposure to sunlight, will serve as the primary energy source. Complemented by nuclear reactors and advanced batteries, the base will have a stable and reliable power supply. Backup systems such as fuel cells will ensure operations continue uninterrupted during long lunar nights or unexpected disruptions.

The Moon's natural resources, such as water ice, will be mined and processed into breathable oxygen, drinkable water, and hydrogen for fuel. This resource utilization will reduce dependency on Earth, making long-term habitation and exploration more feasible. Additionally, the Moon's low gravity and proximity to Earth make it an ideal launchpad for missions to Mars and beyond.

A permanent moon base represents a technological milestone and a profound leap for humanity. It's a statement of our collective ambition, resilience, and ability to adapt to new frontiers. By 2030, the Moon will no longer be a distant dream but an integral part of humanity's journey into the cosmos, bringing us closer to becoming a truly spacefaring civilization.

Meanwhile, asteroid mining unlocks access to resources such as rare earth metals and water, which are critical for Earth's industries and sustainable deep-space exploration. Beyond that, the groundwork is being laid for lunar colonies that will serve as stepping stones for missions to Mars. Interstellar travel, once confined to science fiction, is now being actively developed with breakthroughs in developing sustainable propulsion systems for nuclear propulsion, ion thrusters, and even early concepts for warp drives that can enable long-duration missions to deep space.

Space exploration in 2030 is not just about reaching farther; it's about integrating space into everyday life on Earth, ensuring resources, opportunities, and possibilities are not limited by the boundaries of our atmosphere. It's practical, innovative, and deeply collaborative, setting the stage for humanity's next great leap forward.

WHAT IS GOING TO HAPPEN?

Humanity will reach extraordinary milestones in space exploration, fundamentally transforming life on Earth and beyond. The construction of a permanent moon base will mark a significant leap forward. This base, positioned on the lunar south pole, will be a hub for scientific research, resource mining, and a launchpad for deeper space missions to Mars and beyond. Built with cutting-edge technologies such as 3D printing using lunar regolith to create sustainable habitats, the moon base will symbolize our ability to extend human presence beyond Earth.

Space Trends:

1. **Reusability:** Reusable rockets and spacecraft will continue to reduce the cost of access to space, making it more feasible for commercial ventures to launch payloads and conduct missions.

2. **Artificial Intelligence (AI) and Machine Learning (ML):** AI and ML will be critical in optimizing space missions, improving data analysis, and enhancing decision-making.

3. **3D Printing and Additive Manufacturing:** 3D printing will enable the rapid production of space components, reducing lead times and costs.

4. **In-Orbit Assembly and Manufacturing:** The ability to assemble and manufacture spacecraft and satellites in orbit will revolutionize how we approach space missions and reduce the need for expensive, complex, ground-based infrastructure.

Closer to home, space elevators, once confined to science fiction, will begin to take shape. Made possible by breakthroughs in materials such as carbon nanotubes and graphene, these ropes connecting Earth's surface to geostationary orbit will slash the cost of space transport from $22,000 to just $200 per kilogram. This innovation will democratize access to space, enabling more frequent missions, expanding orbital industries such as satellite maintenance and asteroid mining, and making off-world construction a reality.

Back on Earth, hypersonic transport will revolutionize global logistics and travel, enabling goods and passengers to reach any point on the planet in under two hours. Reusable rockets and suborbital flight technologies, capable of flying at speeds exceeding Mach 5, will reduce delivery times for critical goods such as medical supplies and enhance ultra-fast commercial travel. Companies such as SpaceX and Blue Origin are already paving the way, though challenges in airspace regulations and environmental impacts remain.

Meanwhile, space-based solar power (SBSP) will redefine how we think about energy. Massive orbital solar arrays will capture sunlight 24/7, free from atmospheric interference, and beam it back to Earth using wireless transmission. This continuous and sustainable energy source will reduce fossil fuel dependency, power remote areas, and support space stations and lunar bases. Projects by nations such as China, the U.S., and Japan are leading this ambitious effort.

For deeper exploration, ion propulsion systems will take center stage. Far more efficient than chemical rockets, these engines will provide continuous thrust for long-duration space missions. By 2030, ion propulsion could power missions to Mars, exploratory probes to the outer planets, and even lay the groundwork for interstellar travel. With future iterations potentially powered by nuclear reactors or fusion, humanity will take its first steps toward exploring neighboring star systems.

These advancements will integrate space into our daily lives and lay the foundation for a spacefaring civilization, opening up boundless opportunities for exploration, innovation, and sustainability. The year 2030 will mark the dawn of a new era, where the stars are no longer out of reach, and the future is being built on Earth and in the cosmos.

WHY DOES IT MATTER?

These advancements will fundamentally transform human civilization, redefining how we live, work, and interact with our planet and beyond. The space elevator, for instance, will revolutionize access to orbit by dramatically reducing the cost of transporting goods and people. What currently costs around $22,000 per kilogram could drop to an astonishing $200 per kilogram, making space accessible to more nations, businesses, and individuals. This affordability will pave the way for off-planet industries, such as microgravity manufacturing, which could produce materials like ultra-pure crystals and pharmaceuticals that are impossible to create on Earth, and asteroid mining, which could unlock rare resources critical for technology and clean energy.

Meanwhile, solar power beaming from space will offer a game-changing solution to the energy crisis. By capturing sunlight in orbit—uninterrupted by weather or nightfall—and transmitting it to Earth via wireless energy beams, we'll gain a reliable, clean energy source that could drastically reduce reliance on fossil fuels, cutting emissions and playing a crucial role in combating climate change.

The potential of SBSP to transform global energy systems is immense. Unlike traditional fossil fuels, which deplete finite resources and contribute to climate change, space-based solar power offers a renewable solution with no greenhouse gas emissions. Unlike terrestrial solar panels, which are limited by geography and weather, SBSP can deliver energy continuously, day and night, rain or shine.

The applications are far-reaching. Remote and off-grid areas could receive reliable electricity, revolutionizing energy access for underserved regions. Urban centers could transition away from fossil fuels more rapidly, accelerating the shift to carbon-neutral economies. Future space stations and lunar or Martian habitats could rely on SBSP, making it a cornerstone of space exploration infrastructure.

While challenges remain—such as the cost of launching massive solar arrays into orbit and ensuring the safety of microwave energy transmission—advances in rocket reusability, lightweight materials, and wireless power transmission are rapidly bringing SBSP closer to reality. Nations like China, the U.S., and Japan are already investing in large-scale projects, aiming to deploy operational systems by 2030.

Space-based solar power isn't just a technological innovation—it's a lifeline for a planet facing rising energy demands and environmental challenges. By harnessing the sun's power from above, SBSP has the potential to light up the Earth in ways that are both sustainable and transformative.

These innovations won't just impact the economy—they will democratize access to space, enabling countries, companies, and even individuals to contribute to the exploration and utilization of space. Beyond the tangible benefits, they symbolize humanity's next great leap, inspiring generations to dream bigger and push boundaries while addressing pressing global challenges such as sustainability, resource scarcity, and energy security. These technologies matter because they can transform Earth into a more equitable, innovative, and sustainable home while opening the door to limitless possibilities in space.

WHO WILL IT IMPACT?

The new space economy will profoundly impact everyone, from governments and industries to everyday citizens, reshaping life on Earth and beyond. Governments will play pivotal roles, with nations such as Japan, UAE, and India emerging as space exploration and innovation leaders. These advancements will foster international collaboration and redefine geopolitical and economic balances as access to space becomes a cornerstone of national progress and security.

Global spaceports will emerge as critical hubs, enabling the launch of satellites, cargo, and passengers into orbit. These spaceports, strategically located worldwide, will function like today's airports, creating new avenues for trade, tourism, and technology. Cities near these ports will experience economic booms, attracting businesses, researchers, and travelers eager to explore the cosmos.

Private industries will benefit enormously. Mining companies will look to asteroids as sources of rare materials, while manufacturers will utilize microgravity to produce superior products. Energy companies will harness space-based solar power, transmitting clean energy back to Earth and potentially replacing fossil fuels. Space tourism will no longer be a luxury for the elite as costs decrease, allowing a broader population to experience the wonders of space.

Asteroid Mining and Lunar Colonies

As humanity continues to push the boundaries of space exploration and development, two exciting and interconnected endeavours are gaining momentum: asteroid mining and lunar colonies. These initiatives are poised to revolutionize our access to resources, pave the way for sustainable space travel, and lay the groundwork for humanity's next great leap: establishing a presence on Mars.

Asteroid mining, also known as asteroid prospecting or asteroid resource utilization, is extracting valuable resources from asteroids. These small, rocky bodies are remnants from the solar system's early days, and they hold the key to unlocking a new era of space exploration and development.

Asteroids are rich in rare earth metals, a group of 17 elements essential for many modern technologies, including electronics, renewable energy systems, and advanced ceramics. These metals are found in limited quantities on Earth, making asteroids an attractive supply source. By mining asteroids, we can access these critical resources, reducing our reliance on Earth-based supplies and supporting the growth of sustainable industries

Lunar colonies will provide a testing ground for technologies and strategies used on Mars, reducing the risks and costs associated with establishing a human presence on the Red Planet. Lunar colonies will demonstrate the feasibility of sustainable development in space, paving the way for future human settlements on Mars and beyond.

For everyday citizens, the effects will be transformative. Developing space elevators and affordable transport systems will create faster, cheaper, and more sustainable logistics, improving global trade and infrastructure. Space technologies will spur innovations in health, materials, and sustainability, unexpectedly enhancing life on Earth.

The rise of global spaceports and the expanding space economy will create new opportunities for exploration and innovation and symbolize humanity's transition into a truly spacefaring civilization where all share the benefits of the cosmos.

National Space Agencies

National space agencies are at the forefront of space exploration and research, shaping their countries' roles in the global space economy. They oversee planning, funding, and executing missions while providing a framework for private companies to thrive.

- **NASA (United States):** With a budget exceeding $20 billion, NASA leads human spaceflight, robotic exploration, and groundbreaking space science, maintaining its reputation as a global leader in space innovation.

- **Roscosmos (Russia):** Russia's federal space agency, operating with over $2 billion, focuses on human spaceflight, satellite development, and space science, continuing its legacy in space exploration.

- **ESA (Europe):** The European Space Agency, with a $5 billion budget, coordinates space activities across Europe, excelling in human spaceflight, robotic missions, and scientific research.

- **China:** Investing over $10 billion annually, China is rapidly advancing in lunar exploration, space station development, and satellite technology, solidifying its position as a significant space power.

- **India:** With a budget of over $1 billion, India has achieved remarkable milestones in lunar missions, satellite technology, and space exploration, emerging as a key player in the global space sector.

- **United Arab Emirates (UAE):** The UAE's $500 million investment in space tourism, lunar exploration, and satellite development underscores its ambition to become a significant force in the industry.

Global Spaceports:

Here's a list of notable spaceports around the world, each serving as a critical hub for launching spacecraft and advancing space exploration:

1. **Spaceport America (New Mexico, USA):**
   - The first and currently the only operational commercial spaceport dedicated to private space travel and exploration.
   - Host companies focused on suborbital flights and commercial space tourism.

2. **Guiana Space Centre (French Guiana):**
   - A significant launch site for the European Space Agency (ESA) and international partners.
   - Ideal equatorial location for launching geostationary satellites.

3. **Baikonur Cosmodrome (Kazakhstan):**
   - It is one of the oldest and largest spaceports globally.

- Operated by Russia, it remains a pivotal site for manned and uncrewed space missions.

4. **Tanegashima Space Centre (Japan):**
   - JAXA manages Japan's primary launch site.
   - Known for its advanced technology and contribution to satellite launches.

5. **Kennedy Space Center (Florida, USA):**
   - A cornerstone of NASA's space exploration, hosting historic missions and current commercial launches.
   - A hub for reusable rocket technology and lunar missions.

6. **Wenchang Spacecraft Launch Site (China):**
   - It supports China's ambitious space program in Hainan Province, including satellite launches and lunar missions.

7. **Satish Dhawan Space Centre (India):**
   - Operated by ISRO, it is a critical launch site for India's satellite and lunar missions.
   - Plays a key role in India's growing space capabilities.

8. **Mohammed bin Rashid Space Centre (UAE):**
   - Central to the UAE's space ambitions, focusing on satellite development, Mars exploration, and space tourism.

9. **Alcantara Launch Center (Brazil):**
   - Positioned near the equator, it is ideal for launching geostationary satellites.
   - Set to expand operations with international partnerships.

10. Kourou Spaceport Expansion (French Guiana):

- Future developments are underway to expand its capabilities for commercial and international launches.

CALL TO ACTION

Investors, governments, and business leaders must align their efforts to unlock the vast potential of space technologies, including space tourism, space-based solar power, SmallSats, lunar bases, and asteroid mining. Governments should provide funding and resources to accelerate breakthroughs in reusable launch systems and sustainable space infrastructure. Private companies must take charge of innovating cost-effective solutions, from developing advanced satellites for global communication to enabling safe and affordable access to space for tourism and resource extraction.

At the same time, public–private partnerships will be essential to share expertise and resources, fostering rapid progress in fields such as lunar construction and asteroid mining. Regulatory frameworks must be developed to ensure the space economy grows responsibly, balancing innovation with sustainability and global accessibility. By taking these actions today, stakeholders can create a thriving, interconnected space industry that transforms life on Earth and beyond by 2030.

Governments should focus on building global spaceports—hubs for launching rockets, hosting space tourists, and supporting industries like asteroid mining. These spaceports could create thousands of jobs and drive local economies like airports today.

Businesses must invest in space-based renewable energy projects, such as solar power beamed from satellites, which can provide clean, continuous energy to Earth. Imagine your home powered by sunlight collected 24/7 in orbit—this is no longer science fiction but a reality within reach.

On a personal level, individuals should look into career paths in the space sector, whether that's in engineering, environmental science, or even hospitality, as space tourism grows. Schools and communities can set up workshops and events to inspire the next generation of space innovators.

Career Paths in the Space Sector

The space sector offers a various career paths, from traditional roles like aerospace engineering and astrophysics to emerging fields such as space law and space policy. Some of the most in-demand careers in the space sector include:

- **Aerospace Engineers:** Design and develop spacecraft, satellites, and missiles.

- **Data Scientists:** Analyze and interpret large datasets to gain insights into space-related phenomena.

- **Software Developers:** Create software for space-related applications, such as navigation and communication systems.

- **Space Policy Analysts:** Develop and implement space exploration and utilization policies.

- **Space Law Attorneys:** Advise clients on legal issues related to space exploration and utilization.

- **Planetary Scientists:** Study the formation and evolution of planets and other celestial bodies.

- **Astrophysicists** Study the physical nature of celestial objects and phenomena.

- **Space Operations Managers:** Oversee the daily operations of space-based systems and missions.

History of Humanity, Disease, Healthcare and Wellbeing

"Never again will there be in it an infant who lives but a few days, or an old man who does not live out his years; the one who dies at a hundred will be thought a mere child; the one who fails to reach a hundred will be considered accursed." —Isaiah 65:20

"Healthcare at its best is a conversation between a doctor and a patient trying to solve a problem. WHAT PATIENTS WANT... To live their lives with minimal impact of illness... To get support and a sense of control. WHAT DOCTORS WANT...Practice medicine without hassle factor ... Innovation that makes a difference." —Jordan Shlain, MD*

As told in the book *Breakthrough: Elizabeth Hughes, the Discovery of Insulin, and the Making of a Medical Miracle* by Thea Cooper and Arthur Ginsberg, in 1919, 11-year-old Elizabeth Hughes was diagnosed with Type 1 diabetes (T1D), a death sentence at the time. The disease prevented her body from producing insulin, leading to high blood sugar levels. The only available treatment, Dr. Allen's starvation therapy, limited patients to just 400 calories a day, prolonging life but at the cost of severe suffering.

Dr. Allen urged Elizabeth's parents to try and keep her alive, as a cure seemed close. Despite their concerns about the harsh treatment, they committed to his care. Elizabeth endured the agonizing regimen, her body wasting away to just 45 pounds. Her parents watched helplessly as her health deteriorated, debating whether mere survival was worth the suffering. But just as her life seemed to flicker, a miraculous breakthrough in Canada changed everything. Frederick Banting, a

---

* Used with permission

struggling doctor, and Charles Best, his student assistant, had isolated a hormone called insulin from the animal pancreas. After months of trials, debates, and failures, their experiment succeeded, leading to the first-ever insulin injection in 1922. The day Elizabeth Hughes received her first dose, she was on the brink of death—emaciated, malnourished, and too weak to walk without help. However, within hours of receiving the injection, her blood sugar stabilized, and she could begin to eat a regular diet. Elizabeth lived a whole life, passing away in 1981 at 73 years old.

Another child, Teddy Ryder, also benefited from the discovery of insulin. Initially turned away due to limited supplies, Teddy's uncle pleaded with Banting, explaining the boy wouldn't survive until more insulin became available. Banting ultimately provided the treatment, saving Teddy's life. A year later, a healthy and cheerful Teddy wrote to Banting, proudly declaring himself a "fat boy." Teddy lived over 70 years with insulin, passing away at age 76.

These stories illustrate the life-saving impact of insulin's discovery, transforming diabetes from a fatal condition into a manageable disease and offering hope to millions.

By early 1923, more than 1000 diabetics were treated by over 250 physicians in 60 clinics across the United States and Canada.

In 1922, it is estimated that ~300,000 people developed T1D globally, based on an incidence rate of 15 per 100,000 people per year. The life expectancy for children with TD1 before 1922 was typically only a matter of months without insulin.

Through history, 15 million lives have been saved by insulin.

Banting and Best's discovery of insulin wasn't just about science—it was about human perseverance, a race against time to save lives. The struggle wasn't only in the laboratory but also convincing pharmaceutical companies and physicians to believe in the therapy. Eventually, Eli Lilly, an up-and-coming pharmaceutical company, partnered with Banting to produce insulin on a large scale. The effort saved countless lives, and today, insulin is a standard treatment for TD1 worldwide.

This breakthrough was nothing short of a medical miracle but was also a reminder of the limits of human innovation at the time. Unfortunately, because there is no cure, just treatment, in 2023, an estimated ~9 million people globally live with T1D, with ~1.2 million new cases per year and.

Fast forward to the 2020s, and another revolution is on the horizon. With over 9,000 diseases still without an FDA-approved therapy, as Dr. David Fajgenbaum highlights in *Every Cure: Chasing My Cure*, the hope for a cure for these diseases could be helped with AI. Just as the discovery of insulin changed the lives of people with diabetes forever, AI could transform medicine, offering solutions to some of humanity's most pressing health challenges. These include rare genetic disorders, forms of cancer, autoimmune diseases, and neurological conditions that devastate patients and families alike.

As insulin was a game-changer for diabetes, AI can become a lifeline for untreatable diseases. Dr. Atul Gawande points out, "13,600 diagnoses or ways in which the human body can fail. Now more than 6,000 drugs can be prescribed, and 4,000 medical and surgical procedures can be performed."—a massive toolkit, but still far from addressing every disease. AI has the potential to leverage this growing body of knowledge and predict promising treatments for rare and complex conditions that currently baffle researchers and doctors alike.

HISTORY OF DISEASE, HEALTHCARE and WELLBEING

Inspirational Quotes
- "Never again will there be in it an infant who lives but a few days..." - Isaiah 65:20
- "A single death is a tragedy..." - Joseph Stalin

Avoidable Deaths (Stories)
- Elizabeth Hughes & Insulin
  - Diagnosed with Type 1 diabetes in 1919
  - Endured starvation therapy before insulin discovery
  - Received first insulin injection in 1922, lived a full life
- Teddy Ryder
  - Also saved by insulin in the 1920s
  - Lived 70+ years
- Modern Era
  - ~9 million people live with T1D globally (2023)
  - Challenges remain with global access and treatments

Progress in Healthcare
- Life Expectancy
  - Significant improvements over centuries
  - Reduction in infant mortality and infectious diseases
- Key Innovations
  - Vaccines, synthetic fertilizers, sanitation systems
  - AI potential in drug discovery

Statistics
- Global Deaths
  - 62M deaths in 2024, projected to rise to 120M by 2080
  - Leading causes: heart disease, diabetes, cancer
- Disparities
  - Life expectancy gaps between ethnic groups and regions

History of Humanity, Disease, Healthcare, and Wellbeing

Case Studies in Avoidable Deaths
- Suicide
  - 700,000 annual deaths globally
  - Linked to loneliness, mental health issues
- Obesity
  - 5M deaths annually
  - 1B obese by 2030 globally
- Falls
  - 684,000 fatal falls annually worldwide
  - Prevention strategies: medication reviews, balance training

Technologies That Saved Billions
- Toilets, vaccines, blood transfusions, synthetic fertilizers
- Green Revolution and sanitation innovations

Disability and DALYs
- 1B people globally live with disabilities (15%)
- Major contributors: non-communicable diseases, injuries, mental disorders

Future Challenges and Opportunities
- AI in Healthcare
  - Drug discovery, predictive diagnostics
  - Potential to address 9,000+ untreated diseases
- Call to Action
  - Focus on diet, exercise, mental health
  - Use AI and technology ethically and effectively

PREDICTIONS

Will you die this year or have some primary health concern?

According to the Social Security Actuary Table (https://www.ssa.gov/oact/STATS/table4c6.html), a typical 60-year-old male/female has a ~1.4%/0.87% chance of dying in the year; a 40-year-old male has 0.38%/0.21% chance of dying, or is 3.7/4.1 times less likely to die than a 60-year-old. An 80-year-old male/female has 6.5%/4.6% chance of dying or 4.6/5.3 times more likely to die than a 60-year-old.

See the table that outlines the probability of death in the year, the relative risk likelihood of dying based on a 60-year-old, and life expectancy at 10, 40, 60, 80, and 100.

Age Group	Male Death Probability	Female Death Probability	Relative Likelihood (Male)	Relative Likelihood (Female)	Life Expectancy (Male)	Life Expectancy (Female)
10 years	0.0129%	0.0103%	108.5x < 60	84.5x < 60	74.1	79.8
40 years	0.38%	0.21%	3.7x < 60	4.1x < 60	76.6	81
60 years	1.4%	0.87%	Baseline	Baseline	80.4	83.6
80 years	6.5%	4.6%	4.6x > 60	5.3x > 60	87.9	89.4
100 years	36%	32%	2,781x >10 year old	3,074x >10 year old	102	102.4

We are shocked when a child dies and we expect a 100-year-old to die very soon. That is because a child is ~3,000x less likely to die in the year. When we are young, we think we are immortal, and do many stupid things. As we age, we understand we are mortal.

WHY DOES IT MATTER?

Understanding humanity's health journey highlights the value of innovation, the resilience of the human spirit, and progress. Advances in extending life expectancy and reducing preventable deaths transform potential into reality. It matters because each step forward fosters hope, improves quality of life, and builds a foundation for a healthier future for all.

CALL TO ACTION

What problems are you trying to solve? For many, especially as they age, one of the most significant problems is health; finally, it is the only issue they are trying to solve.

Here are some top items to focus on to avoid death and maximize your healthy lifespan: diet, exercise, minimizing reckless activities, having healthy friendships with people who love and care for you and you for them, and having access to good quality healthcare following your treatment plan.

This could be summed up with the saying, "Don't do drugs, drink, drive distracted or speed, smoke, overeat or chew or go out with reckless people that do."

Or, more positively, "Eat well, stay fit, keep friends who don't sit, and go out with healthy friends who do not let you quit."

However, it is essential to avoid ending up like Tithonus. He was a handsome mortal, so naturally, the goddess Eos (the Dawn) fell head over heels for him. Like anyone obsessed, she wanted to keep him around forever. So, she went to Zeus, king of the gods, and requested a little favour: "Grant my beloved Tithonus eternal life." Zeus, always down for a laugh, agreed. Tithonus would never die. But one small catch—Eos forgot to ask for eternal youth. As the years rolled by, Tithonus aged. And age and aged. Imagine becoming a wrinkled, toothless husk, desperately shuffling through eternity. Eventually, the gods took pity on him and turned him into a cicada. Immortal, yes—but youthful and happy? Not so much

DOUG HOHULIN ON USING AI TO SAVE ONE BILLION LIVES

In 2022, I was the project manager for the MediView 10G: The Best Breast Cancer Care Everywhere project led by Dr. Jamie Wagner, that used AI and XR/Spatial Computing using the HoloLens 2 to help patients in rural areas access care from a National Cancer Institute (NCI)-designated Cancer Center.

A video from CableLabs on the future of aging states: "Technology keeps us connected and engaged with new experiences. As innovations push our health expectancy as far out as our life expectancy, technology offers an independence that has been hard to achieve for the aging population. It's not what we create but why. What is even more incredible than the future tech is what the creations make possible."

According to World bank's Disability Inclusion Overview, this technology can benefit the "one billion people, or 15% of the world's population, experience some form of disability."

100 Billion Humans: A History of Disease

It has been estimated that 100 billion humans have lived on the earth throughout history. For the 92 billion people that have already died, the reasons have varied significantly over time, with different causes of death prevalent in different eras. Below are some of the significant causes across human history:

1. Childbirth and Infant Mortality

- Historical infant mortality: 40–50% of children died before the age of 5 (pre-modern medicine).

- Maternal deaths in childbirth: Tens of millions over millennia. The maternal mortality rate before 1800 is estimated to have been between 1% and 1.5% per birth. If a woman had 6–7 children, the cumulative risk could be 6–10% over her reproductive years.

2. Violence and Warfare

- Wars:
    - World War I: 15–20 million deaths.
    - World War II: 60–85 million deaths.
    - Other historical conflicts (Crusades, Mongol invasions, etc.): Tens of millions.

- Genocides and Massacres:
    - Holocaust: 6 million deaths.
    - Mongol conquests: Estimated 30–40 million deaths.

- Homicides and Crime: Tens of millions over history.

- Per-War Risk: During significant conflicts, the death probability for soldiers in battle could range from 5% to 20%. This depends on the scale and lethality of the war, with higher risks in significant conflicts such as the Napoleonic Wars or the Thirty Years' War.

- Thirty Years' War (1618–1648): In some regions of Europe, up to 20–30% of the male population died, including war-related famine and disease.

- Napoleonic Wars (1803–1815): Soldiers had a 15–25% chance of dying in battle or due to war-related causes.

- Cumulative Lifetime Risk: When considering a man's lifetime up to 1900, factoring in periods of peace and conflict, the lifetime risk of dying in war could be estimated at 2–5% on average across most societies, depending on the country and its history of warfare.

3. Infectious Diseases

- Plagues and Epidemics:
  - Black Death (bubonic plague): 75–200 million deaths.
  - Smallpox: 300–500 million deaths (over centuries).
  - Spanish flu: 50–100 million deaths (1918–1919).

- Chronic Infectious Diseases:
  - Tuberculosis: 1 billion deaths (over millennia).
  - Malaria: Estimated 300–500 million deaths.
  - Cholera: Millions over multiple outbreaks.
  - Influenza (various strains): Tens of millions over centuries.

4. Malnutrition and Starvation

- Famine-related deaths: Hundreds of millions.
  - Chinese famine (1958–1962): 15–45 million deaths.
  - Irish Potato Famine: 1 million deaths.

5. Natural Disasters

- Major disasters (earthquakes, tsunamis, floods, etc.): Tens of millions.
  - 1931 China floods: Estimated 1–4 million deaths.
  - 2004 Indian Ocean tsunami: 230,000–280,000 deaths.

6. Aging and Degenerative Diseases

- Modern diseases such as heart disease, cancer, and strokes: Hundreds of millions (particularly in the last century).

7. Poor Sanitation and Unsafe Water

- Cholera and dysentery (due to contaminated water): Tens of millions over millennia.

8. Environmental Exposure

- Cold, heat, and exposure to elements (prehistoric/early human deaths): Tens of millions.

9. Accidents

- Accidental deaths (drowning, falls, fires, animal attacks, etc.): Hundreds of millions over human history.

10. Hunting and Gathering Hazards

- Early human deaths due to hunting risks: Likely millions.

11. Toxic Substances and Poisoning

- Deaths from food poisoning, venomous bites, and toxic plants: Millions over history.

As highlighted in the video "How They Did It - Growing Up Roman" the probability of death by age during Roman times started at 40–50% in a single year and was at the lowest when you were 10 years old at ~8%. At 10, life expectancy was ~12.5 additional years on average, indicating that reaching old age was uncommon, and many individuals did not survive past their early twenties.

COVID PANDEMIC AND LIFE EXPECTANCY

Unfortunately, pandemics can arise that decrease life expectancy. The article "Ten Americas: a systematic analysis of life expectancy disparities in the USA" (Dwyer-Lindgren, Laura et al., *The Lancet* 404.10469, 2299-2313, https://www.thelancet.com/journals/lancet/article/PIIS0140-6736(24)01495-8/fulltext) includes ten groups based on race, ethnicity, geography, and socioeconomic factors. It highlights growing inequalities in life expectancy in the U.S. from 2000 to 2021, with gaps increasing from 12.6 years in 2000 to 20.4 years in 2021. Significant declines occurred during COVID-19, particularly for American Indian and Alaska Native populations. While some groups, notably Black Americans, showed progress pre-202,0, others have not, emphasizing the need for targeted policies addressing health and socioeconomic disparities.

Group 10, "AIAN | West (American Indian or Alaska Native,)" has a 13.6-year lower life expectancy than Group 3, "White majority, Asian, AIAN | Other counties," and 20.4-year lower life expectancy than Group 1, "Asian."

Group	2000 (years)	2019 (years)	2021 (years)	2021 -2019 Decline	Delta from Group 3
1 Asian	83.1	86	84	-2.0	6.8
2 Latino | Other counties	80.4	83	79.4	-3.6	2.2
3 White majority, Asian, AIAN | Other counties	77.5	79.3	77.2	-2.1	Ref
4 White | Non-metropolitan and low-income Northlands	77.6	78.6	76.7	-1.9	-0.5
5 Latino | Southwest	77.8	80.4	76	-4.4	-1.2
6 Black | Other counties	72	75.7	72.3	-3.4	-4.9
7 Black | Highly segregated metropolitan areas	70.6	74.9	71.5	-3.4	-5.7
8 White | Low-income Appalachia and Lower Mississippi Valley	74.8	74.8	71.1	-3.7	-6.1
9 Black | Non-metropolitan and low-income South	70.5	72.5	68	-4.5	-9.2
10 AIAN | West (American Indian or Alaska Native).	72.3	70.2	63.6	-6.6	-13.6

"Ten Americas: A Systematic Analysis of Life Expectancy Disparities in the USA" includes the following key calls to action to address health disparities:

1. **Reduce Health Inequities:** Prioritize interventions to address systemic barriers contributing to health disparities, including socioeconomic inequality, racial and ethnic discrimination, and geographic disparities.

2. **Focus on Marginalized Groups:** Design targeted strategies for the most disadvantaged groups, particularly those with the lowest life expectancy (e.g., AIAN populations and low-income rural communities).

3. **Improve Data Collection**: Invest in more comprehensive and accurate data collection to understand and address the nuanced interplay of factors influencing health outcomes.

4. **Enhance Equity in Healthcare Access**: Expand equitable access to quality healthcare, preventive measures, and treatment, particularly in underserved areas.

5. **Invest in Socioeconomic Factors**: Address social determinants of health, such as income, education, and employment opportunities, to reduce disparities in life expectancy.

6. **Leverage Local and National Planning**: Use data-driven insights to guide local initiatives and national policies to combat the root causes of poor health outcomes.

7. **Monitor and Evaluate Progress**: Establish mechanisms to regularly assess the effectiveness of interventions to reduce health inequities and adapt strategies as needed.

8. **Address COVID-19 Impacts**: Develop specific recovery strategies to mitigate the disproportionate effects of the COVID-19 pandemic on vulnerable populations.

These calls to action underscore the urgent need for multi-faceted approaches to reduce disparities and promote health across all communities in the U.S.

Humanity's Legacy and Lessons from 100 Billion Lives

The journey of humanity, represented by the estimated 100 billion individuals who have ever lived, offers profound insights into resilience, survival, and adaptation. Examining the causes of death across eras reveals not only the vulnerabilities of our ancestors but also the tremendous progress made in modern times. From the staggering infant mortality rates of pre-modern societies to the devastating tolls of wars, plagues, and famine, human history has been shaped by both external forces and the ingenuity to overcome them.

In earlier epochs, childbirth, infectious diseases, and environmental hazards were formidable challenges. Maternal and infant mortality rates underscore the fragility of life before the advent of modern medicine. Similarly, historical conflicts—whether through warfare or genocide—claimed millions of lives, serving as stark reminders of the consequences of unchecked violence and political strife. Epidemics such as the Black Death and smallpox left indelible marks on the human population, but they also catalyzed advancements in public health and sanitation. Over time, the shift from acute infectious diseases to chronic degenerative conditions such as heart disease and cancer reflects a transformation in human longevity and lifestyle. Yet, the persistence of preventable deaths from malnutrition, poor sanitation, and unsafe water highlights the disparities that remain.

Today, humanity enjoys more safety and longevity. The probability of death, once alarmingly high at every life stage, has plummeted for modern individuals. Advances in healthcare, technology, and societal structure allow most people to envision lifespans far exceeding those of their ancestors. However, this progress brings new challenges: aging populations, lifestyle diseases, and the existential threats of pandemics and global conflicts.

Top Causes of Death

As outlined by the WHO in "The Top 10 Causes of Death", in 2021, the top 10 causes of death accounted for 39 million deaths, or 57% of the total 68 million deaths globally (higher than 2023 due to Covid). Noncommunicable diseases, primarily cardiovascular and respiratory diseases, make up 68% of these deaths, highlighting the global shift toward chronic health conditions. The top causes include:

1. **Ischaemic heart disease**: 9.1 million deaths (~13% of total deaths)

2. **COVID-19**: 8.8 million deaths

3. **Stroke**: 6.6 million deaths

4. **Chronic obstructive pulmonary disease (COPD)**: 3.23 million deaths

5. **Lower respiratory infections**: 2.5 million deaths

6. **Trachea, bronchus, and lung cancers**: 1.9 million deaths

7. **Alzheimer's disease and other dementias**: 1.8 million deaths

8. **Diabetes**: 2 million deaths, with a 95% increase since 2000

9. **Kidney diseases:** 2.4 million deaths

10. **Liver cirrhosis**: 1.3 million deaths

Despite technological advances, millions continue to die due to diseases that are preventable or manageable with timely intervention.

> "In 2022, approximately 20 million cancer cases were newly diagnosed, and 9.7 million people died from the disease worldwide. By 2050, the number of cancer cases is predicted to increase to 35 million based solely on projected population growth." Global Cancer Facts & Figures, 5th edition. Atlanta: American Cancer Society, Inc. 2024.

Case Study: Avoiding Avoidable Deaths - Suicide

Suicide death should be one of the most avoidable deaths to avoid. "More than 700,000 persons die by suicide every year globally. Suicide is the fourth leading cause of death among 15-29 year olds." (WHO Suicide Worldwide in 2019).

"In 2022, suicide was among the top 9 leading causes of death for people ages 10-64. Suicide was the second leading cause of death for people ages 10-14 and 25-34. Suicide rates increased approximately 36% between 2000–2022. Suicide was responsible for 49,476 deaths in 2022, which is about one death every 11 minutes. The number of people who think about or attempt suicide is even higher. In 2022, an estimated 13.2 million adults seriously thought about suicide, 3.8 million planned a suicide attempt, and 1.6 million attempted suicide." "Facts About Suicide: Suicide Prevention", Centers for Disease Control and Prevention

These factors highlight the reduction of lifespan that contributes to suicide.

1. **Loneliness:** Associated with a 26–32% increased risk of early mortality, equating to several lost years, comparable to smoking

2. **Smoking:** Reduces lifespan by about ten years on average

3. **Opioid Addiction:** Reduces life expectancy significantly, with overdose risks and related health issues cutting 10–15 years of life.

4. **Heavy Alcohol Use:** Can reduce life expectancy by up to 20 years for those with chronic alcohol use disorders

5. **Low Physical Activity:** Sedentary individuals may lose 3–5 years of life compared to regularly active ones

6. **Depression:** Can shorten lifespan by about 5–8 years, especially if untreated.

7. **Chronic Stress:** Linked to higher risks of heart disease and other illnesses, reducing life expectancy by up to 3–5 years.

8. **Sleep Disorders:** Chronic sleep deprivation or untreated sleep apnea can reduce lifespan by 3–5 years

WISQARS LEADING CAUSES OF DEATH VISUALIZATION TOOL SHOWS THE IMPACT OF SUICIDE BY AGE GROUP.

10 Leading Causes of Death, United States
2022, All Deaths with drilldown to ICD codes, All Sexes, All Races, All Ethnicities

■ Unintentional Injury ■ Homicide ■ Suicide

	<1	1-4	5-9	10-14	15-24	25-34	35-44	45-54	55-64	65+	All Ages
1	Congenital Anomalies 3,970	Unintentional Injury 1,288	Unintentional Injury 726	Unintentional Injury 926	Unintentional Injury 14,669	Unintentional Injury 33,058	Unintentional Injury 36,872	Malignant Neoplasms 33,363	Malignant Neoplasms 105,133	Heart Disease 557,365	Heart Disease 702,880
2	Short Gestation 2,884	Congenital Anomalies 441	Malignant Neoplasms 393	Suicide 493	Homicide 6,262	Suicide 8,663	Heart Disease 12,258	Heart Disease 32,298	Heart Disease 85,733	Malignant Neoplasms 452,490	Malignant Neoplasms 608,371
3	Sids 1,529	Homicide 343	Congenital Anomalies 241	Malignant Neoplasms 442	Suicide 6,040	Homicide 6,712	Malignant Neoplasms 11,177	Unintentional Injury 31,384	Unintentional Injury 34,017	Covid-19 146,320	Unintentional Injury 227,039
4	Unintentional Injury 1,354	Malignant Neoplasms 296	Homicide 180	Homicide 366	Malignant Neoplasms 1,421	Heart Disease 3,789	Suicide 8,185	Covid-19 9,676	Covid-19 24,252	Cerebrovascular 142,513	Covid-19 186,552
5	Maternal Pregnancy Comp. 1,215	Influenza & Pneumonia 129	Influenza & Pneumonia 77	Congenital Anomalies 205	Heart Disease 848	Malignant Neoplasms 3,641	Liver Disease 5,501	Liver Disease 9,401	Diabetes Mellitus 17,410	Chronic Low. Respiratory Disease 125,803	Cerebrovascular 165,393
6	Placenta Cord Membranes 849	Heart Disease 103	Heart Disease 73	Heart Disease 145	Covid-19 447	Liver Disease 1,786	Homicide 4,765	Suicide 7,781	Chronic Low. Respiratory Disease 17,138	Alzheimer's Disease 118,525	Chronic Low. Respiratory Disease 147,382
7	Bacterial Sepsis 636	Covid-19 101	Covid-19 62	Covid-19 69	Congenital Anomalies 412	Covid-19 1,640	Covid-19 3,841	Diabetes Mellitus 7,364	Liver Disease 16,484	Unintentional Injury 72,816	Alzheimer's Disease 120,122
8	Respiratory Distress 456	Perinatal Period 62	Chronic Low. Respiratory Disease 48	Chronic Low. Respiratory Disease 58	Diabetes Mellitus 324	Diabetes Mellitus 1,188	Diabetes Mellitus 2,879	Cerebrovascular 5,563	Cerebrovascular 14,173	Diabetes Mellitus 71,985	Diabetes Mellitus 101,209
9	Intrauterine Hypoxia 362	Septicemia 60	Cerebrovascular 45	Cerebrovascular 55	Chronic Low. Respiratory Disease 197	Cerebrovascular 599	Cerebrovascular 2,150	Chronic Low. Respiratory Disease 2,987	Suicide 7,864	Nephritis 47,086	Nephritis 57,937
10	Circulatory System Disease 356	Cerebrovascular 49	Septicemia 33	Influenza & Pneumonia 54	Influenza & Pneumonia 188	Complicated Pregnancy 591	Nephritis 1,029	Homicide 2,740	Nephritis 8,668	Parkinson's Disease 38,931	Liver Disease 54,803

In 2024, 61% of couples met online. AI (for better or worse) may reduce loneliness if appropriately used (How Couples Meet? Data from 1930 to 2024).

AI and social media are powerful tools that can be used for good or bad. We need to educate people on using these tools to get the most benefit and minimize the most harm. This will be discussed in the next chapter.

The report "What Our Research Says About Teen Well-Being and Instagram" highlights how social media helps many people, but roughly equal numbers are harmed. AI will likely show similar results (https://about.fb.com/news/2021/09/research-teen-well-being-and-instagram/).

Instagram has a dual impact on users, fostering connections for some while worsening mental health issues for others.

Positive Aspects

- Instagram positively impacts some users' social connections and general experiences with specific stressors. For example, nearly half (45%) felt that Instagram alleviated feelings of loneliness, suggesting that the platform can foster a sense of connection.

- It also benefited a large portion of users dealing with work stress (41.5%) and family stress (43.6%), showing that some individuals find Instagram to be a helpful outlet or distraction from daily pressures.

Negative Aspects

- **Problematic Use:** 31.1% of users reported that Instagram worsened problematic usage patterns, with many users feeling "hooked" or unable to moderate their time spent on the app.

- **Social Comparison and Body Image:** These issues saw a significant negative impact, with 22.7% and 18% feeling that Instagram intensified their concerns around appearance and comparison to others. This highlights the platform's potential role in fueling insecurities through curated content and idealized portrayals of life.

- **FOMO (Fear of Missing Out):** A substantial 18.6% of users experienced heightened FOMO, indicating that Instagram can exacerbate feelings of exclusion or envy when users see others' social activities.

- **Sleep Issues, Suicidal Ideation, and Anxiety:** For these serious concerns, Instagram also had a measurable negative impact. Sleep issues worsened for 17.7% of users, while 15.3% and 14.8%, respectively, noted an increase in suicidal ideation and anxiety.

While Instagram serves as a tool for connection and can reduce certain types of stress and loneliness for a significant number of people, it also presents substantial risks for others, particularly around mental health and self-perception. This dual effect underscores the need for mindful engagement with social media and support structures for users more vulnerable to its adverse impacts.

Case Study 2: Avoiding Avoidable Deaths – Obesity

As outlined in the article " How 'miracle' weight-loss drugs will change the world" models suggest societal upheaval from anti-obesity medicines — but impacts are hard to predict.

> Per Sara Reardon, "of the estimated 5 million deaths each year caused by obesity-associated conditions, 77% occur in low- and middle-income countries" (www.nature.com/articles/d41586-024-03589-7)

The World Obesity Atlas 2022, published by the World Obesity Federation, predicts that one billion people globally, including one in five women and one in seven men, will be living with obesity by 2030. The report shows the global prevalence of obesity across different years (2010, 2020, 2025, and 2030) and categorizes obesity into three classes (Class I, Class II, and Class III).

Year	Class I Obesity	Class II Obesity	Class III Obesity	Total Prevalence (in millions)
2010	368 million	101 million	42 million	511 million
2020	526 million	161 million	77 million	764 million
2025	608 million	191 million	93 million	892 million
2030	692 million	222 million	111 million	1,025 million

The impact of different classes of obesity on mortality varies. (National Institutes of Health):

1. **Class I Obesity (BMI 30.0–34.9):** May reduce life expectancy by 2–5 years.

2. **Class II Obesity (BMI 35.0–39.9):** May result in a 6–10 year reduction in lifespan due to higher rates of heart disease, stroke, and cancer.

3. **Class III Obesity (BMI ≥ 40):** has been found to reduce life expectancy by up to 14 years. Excess deaths are predominantly caused by heart disease, cancer, diabetes, and respiratory issues. The life expectancy reduction in this class is comparable to that seen among heavy smokers.

With the total number of individuals with obesity rising to over one billion by 2030, we can expect substantial increases in the global burden of obesity-related mortality.

Case Study 3: Avoiding Avoidable Deaths – Don't Fall

Over 3 million older adults in the United States visit emergency departments yearly due to falls. In comparison, there are about 2.5 million emergency department visits annually for automobile accident–related injuries. This comparison shows that falls, particularly among older adults, are a significant cause of emergency room visits, even more so than car accidents. In 2022, there were 44,630 deaths due to unintentional falls, which is more than the 42,795 motor vehicle traffic fatalities. The cost of safety features in an average car ranges from $590 to $12,285 for advanced safety packages or individual features. Unfortunately, very little is spent on preventing falls, even though the risk is higher. About 20% of older adults who go to the emergency room for a fall die within a year, underscoring the critical need for effective fall prevention strategies. (CDC).

Globally, falls are a significant public health concern. According to the World Health Organization, falls are the second leading cause of accidental or unintentional injury deaths worldwide. It is estimated that there are 684,000 fatal falls each year, making it a serious global health issue, particularly among older adults.

How can you reduce your risk of falling—especially if you are over 60? As outlined in the video "How to Prevent Falls: 4 Approaches to Ask the Doctor to Try":

- Medication review: reduce medications that increase fall risk (always check with your doctor first).

- Check for blood pressure (BP) sitting and standing. (e.g. sitting systolic BP less than 120).

- Gait, strength, and balance evaluations, often in collaboration with physical therapy.

- Home safety assessment and modification should be done in collaboration with occupational therapy.

PAST TECHNOLOGIES THAT SAVED BILLIONS OF LIVES

In the article "These discoveries saved billions of lives" from the World Economic Forum "Today's infographic from AperionCare highlights the top 50 breakthroughs, ranging from pasteurization to the bifurcated needle, that have helped propel global life expectancy upwards." While many of these innovations have some linkage to the medical realm, there are also breakthroughs in sectors such as energy, sanitation, and agriculture that have helped us lead to longer and healthier lives.

The technologies that have saved one billion lives are as follows:

1. **Toilets** (Invented: 1875)
2. **Synthetic Fertilizers** (Invented: 1909)

3. **Blood Transfusions** (Invented: 1913)
4. **Green Revolution** (Agricultural technology enhancements, Invented: 1945)
5. **Vaccines** (Invented: 1955)

The Impact of Technology on Lives Saved

From the AperionCare Analysis, cumulative lives saved by different technologies and interventions:

- **Surgical Procedures**: 70 million lives saved, with an additional 1 million lives saved annually.
  - Innovations such as bypass surgery, angioplasty, and robotic surgery have revolutionized treatment for cardiovascular diseases.

- **Non-surgical Medical Procedures**: 1.104 billion lives saved, with 7 million saved annually.
  - Key advancements include blood transfusions, cancer screening programs, and radiology tools such as MRI and CT scans.

- **Pharmacology:** 1.457 billion lives saved.
  - The development of antibiotics, vaccines, antimalarial drugs, and insulin has been instrumental in managing infectious diseases and chronic conditions such as diabetes.

- **Preventive Medicine:** 283 million lives saved, plus 7.75 million lives saved yearly.
  - Efforts such as vaccination programs, anti-smoking campaigns, and genetic mapping have had long-lasting health impacts.

- **Food & Water Production/Treatment:** 2.45 billion lives saved, with 1 million saved each year.

- Technologies such as toilets, water chlorination, synthetic fertilizers, and the Green Revolution have drastically reduced waterborne diseases and malnutrition.

- **Environmental Protection:** 2 million lives saved, with 500,000 saved annually.
  - Innovations such as air conditioning and renewable energy reduce the impact of climate-related deaths.

- **Technological Tools:** 3.25 million lives saved, with 804,500 lives saved each year (and rising).
  - Auto safety tools such as seatbelts, airbags, and big data for disaster prediction continue to protect lives.

- **Communication & Data:** 1.1 million lives and 1.1 million saved annually.
  - Technologies such as MOOCs (massive open online courses) and drones have extended healthcare access to remote areas.

The extraordinary advancements highlighted by AperionCare demonstrate the profound impact of human ingenuity on global life expectancy. From the invention of toilets and synthetic fertilizers to the Green Revolution and modern vaccines, each breakthrough represents a milestone in the collective effort to combat disease, hunger, and environmental challenges. These technologies remind us that lifesaving solutions often transcend the medical realm, encompassing agriculture, sanitation, and communication tools.

GLOBAL PHYSICAL DISABILITY

The World Health Organization (WHO) "estimates that over 1 billion people worldwide live with some form of disability, accounting for about 15% of the global population."

"The incidence of chronic diseases in the United States (USA) is steadily increasing, i.e., 60% of adults have one chronic disease, and 40% have more than two chronic diseases, totalling USD 3.3 trillion in annual healthcare costs." A Review of the Role of Artificial Intelligence in Healthcare – Ahmed Al Kuwaiti

Disability-Adjusted Life Years (DALYs) are a metric used to assess the overall disease burden by combining years lost due to premature mortality and years lived with disability. One DALY represents one lost year of healthy life. Over the past 20 years, The Lancet has extensively reported on global DALYs through the Global Burden of Disease (GBD) studies conducted by the Institute for Health Metrics and Evaluation (IHME).

Global DALYs: Current Numbers and Trends

As of the latest data up to 2023, here's an overview of global DALYs based on The Lancet's Global Burden of Disease (GBD) studies:

1. **Total Global DALYs:**

   - Estimation: In 2019, the global total DALYs were approximately 2.8 billion.

   - Growth Trend: Over the past two decades, the total DALYs have increased due to population growth, aging, and epidemiological transitions.

2. **Leading Causes of DALYs:**

   - Non-Communicable Diseases (NCDs):

     - Cardiovascular Diseases: The leading cause, accounting for about 20% of global DALYs.

- Cancer: Approximately 10–12% of DALYs.

- Chronic Respiratory Diseases: Around 5–7%.

- Diabetes Mellitus: Roughly 3–4%.

- Communicable, Maternal, Neonatal, and Nutritional Diseases:

  - Lower Burden Compared to Two Decades Ago: These accounted for about 25% of DALYs, reflecting improvements in infectious disease control but still significant in low-income regions.

- Injuries:

  - Road Traffic Accidents: A major contributor, especially in low- and middle-income countries.

  - Other Injuries: Including occupational injuries, violence, and unintentional injuries, collectively accounting for around 8–10% of DALYs.

- Mental and Substance Use Disorders:

  - Significant and Growing: Conditions like depression and anxiety contribute approximately 5–6% of DALYs globally.

Over the last 20 years, the global number of Years Lived with Disability (YLD) has increased in absolute terms due to demographic changes and the rising prevalence of non-communicable diseases and injuries. While precise annual figures are scarce, estimates indicate a growth from approximately 760 million YLDs in 2000 to about 850 million YLDs in 2019. This trend underscores the need for global health strategies focusing on prevention, management, and support for chronic health conditions that contribute to long-term disability.

Lessons from Humanity's Legacy and the Path Forward

The history of humanity, encompassing the lives of an estimated 100 billion individuals, is a story of resilience, innovation, and profound challenges. Through millennia, humans have faced adversities ranging from infectious diseases and warfare to environmental hazards and aging. Yet, our collective ability to adapt, innovate, and strive for better lives has been the defining narrative.

From the heartbreak of high infant mortality rates in ancient times to the transformative discovery of insulin in the 20th century, we have seen how human ingenuity has repeatedly turned despair into hope. Epidemics such as the Black Death and pandemics such as COVID-19 have served as both tragedies and turning points, driving advancements in medicine and public health. Similarly, groundbreaking innovations such as vaccines, synthetic fertilizers, and sanitation systems have extended lifespans and improved the quality of life for billions.

Yet, the present moment reminds us of the challenges that remain. Preventable deaths due to malnutrition, poor sanitation, and lack of access to healthcare persist in many parts of the world. Chronic conditions such as obesity, heart disease, and diabetes are rising, exacerbated by lifestyle changes and systemic inequities. Meanwhile, mental health issues and loneliness, magnified by modern technology and societal shifts, pose unique 21st-century challenges.

The role of AI, which will be discussed in the next chapter, offers immense potential to address these challenges. The story of humanity is not just about survival—it is about thriving. By learning from the past and leveraging the present innovations, we can build a future where more lives are saved and lived fully, healthily, and meaningfully.

Humanity's Quest to Save a Billion Lives with AI: Avoiding the Most Avoidable Deaths

CURING THE INCURABLE: A NEW ERA BY 2030 WITH SCIENTISTS, CLINICIANS, AND PATIENTS COLLABORATING WITH AI TOOLS

The goal is that by 2030, AI could help identify treatments for hundreds, if not thousands, of currently untreatable diseases. It will accelerate drug discovery and unlock the full potential of precision medicine, matching treatments to patients in ways never before possible. AI also promises to reduce avoidable deaths by addressing the social determinants of health—factors such as poverty, education, and access to care. Predictive models will enable public health officials to target interventions more effectively, ensuring that vaccines, treatments, and health resources reach those in need before it's too late.

The story of Elizabeth Hughes and the discovery of insulin in the earlier chapter reminds us that even in the darkest times, hope can be found through innovation. Just as Banting and Best's discovery gave life to millions of people with diabetes, AI can save millions more by solving the medical mysteries that still elude us.

By 2030, AI could become an essential part of healthcare and wellness care, helping to cure diseases once thought incurable, transforming diagnostics and treatment, and reducing the human suffering caused by preventable deaths. With the power of AI, the dream of curing many of the 9,000 incurable diseases is not just a distant possibility—it is within reach. And just as insulin did for diabetes in the 1920s, AI could give humanity a second chance at life in the 2020s and 2030s. The breakthroughs of the future will echo those of the past, reminding us that the pursuit of cures by dedicated doctors, nurses, caregivers, engineers, and scientists and the will to save lives are what make us truly human. With human innovation using AI tools, we can work towards a day where no patient is left without hope and no disease is left without a solution.

HUMANITIES QUEST TO SAVE A BILLION LIVES IN 2030

Humanities Quest to Save a Billion Lives with AI

Introduction
- AI's potential to save lives.
- Peter Diamandis and Ray Kurzweil on longevity.

AI Predictions
- 1 Million lives saved annually by 2030.
- 10 Million lives saved annually by 2040.
- Disability reduction: 5% by 2030, 25% by 2040.
- ASI Goal: Save 1 Billion lives, 90% disability reduction.

Transformative AI Impact by 2030
- **Cardiovascular Diseases**: Early detection, personalized care; 1.5 million lives saved.
- **Respiratory Diseases**: AI monitoring, treatment optimization; 500,000 lives saved.
- **Cancer Detection**: AI-driven diagnostics; 400,000 lives saved.
- **Diabetes Management**: Predictive AI tools; 200,000 lives saved.
- **Pandemics**: AI for outbreak prediction, vaccine logistics; 1 million lives saved.

Healthcare System Cost Savings
- AI-driven automation and efficiency: $200-$360 billion savings by 2028.
- **Key Areas**: Hospitals, physician groups, private/public payers.

Disability Management with AI
- Advanced prosthetics, diagnostics.
- Predictive analytics for early intervention.
- Assistive tools for accessibility and mobility.

XPRIZE Healthspan Challenge
- Goals: Restore muscle, cognitive, immune functions by 20 years.
- Potential to extend quality life spans significantly.

Reducing Global Inequities with AI
- Address systemic disparities: socioeconomic, racial, geographic.
- AI-driven health monitoring and equitable resource distribution.

Digital Wellbeing in AI Era
- **Six Dimensions of Wellness**: Intellectual, occupational, spiritual, social, emotional, physical.
- Responsible AI design for meaningful connections and personal growth.

Challenges and Risks
- Ethical deployment to prevent AI misuse.
- AI in war and malicious use.

Vision for the Future
- **AI's Role**: Saving lives, extending healthspans, improving accessibility.
- Supporting 10 billion humans sustainably by the 21st century's end.

PREDICTIONS

- 1 million lives saved per year due to AI in 2030

- 1 billion people using AI for healthcare to reduce medical errors, saving 1 million lives a year in 2033

- 10 million lives saved per year due to AI in 2040

- 1 billion people using AI to reduce traffic accidents, saving 1 million lives a year (1.3 million today but may go to 2M before going down) - prediction 2043

- 1 billion lives saved in the 21st century using AI

- 5% reduction in humans with disabilities (50 million people) in 2030 because of AI

- 25% reduction in humans with disabilities (250 million people) in 2040 because of AI

"One billion people, or 15% of the world's population, experience some form of disability" Worldbank.org Disability Inclusion Overview.

What are key questions to consider:

1. Will you die this year or have some primary health concerns?
2. What is your probability of dying this year?
3. How can AI help you avoid death or serious illness today, or in the next 10 days, 1 year, or 10 years?

AGI (Artificial General Intelligence) will arrive for healthcare when AI can save at least 1 million lives in a year, lower the death rate of people under 30 by 50%, reduce the disability rate by 50% and restore muscle, cognitive, and immune function by a minimum of 10 years in persons aged 50–80 years

ASI (Artificial Super Intelligence) will arrive for healthcare when AI can save at least 1 billion lives and reduce the disability rate by 90%, lower the death rate of people under 30 by 90%, and restore muscle, cognitive, and immune function by a minimum of 30 years in persons aged 50–80. One billion people in the global population aged 100 years or older will be doing activities such as skiing, biking 100 miles daily, wake surfing, and hiking 5 miles.

The XPRIZE Foundation aims to restore muscle, cognitive, and immune function by up to 20 years in persons aged 50–80.

> *"The winning team of the $101M XPRIZE Healthspan must demonstrate that their treatment restores muscle, cognitive, and immune function by a minimum of 10 years, with a goal of 20 years, in persons aged 50-80. The treatment must take 1 year or less."*

—XPRIZE Healthspan

Summary

- Reverse age-related degradation by 10–20 years in muscle, cognitive, and immune function

- Single treatment (1-year)

- 7-year timeframe to achieve

- 460 companies are competing for the $101 million Purse.

If the XPRIZE Healthspan goal is achieved, you would lower your risk of dying by 3.7x to 5.3x yearly. How would you value 20 years of high-quality life extension and reduce your risk of dying by 4 times?

As highlighted in the previous chapter, as we age, our risk of dying continues to rise dramatically. Aging is just increasing your risk of dying until you die. Thankfully, a 10-year-old has a very low probability of death, ~100x lower than a 60-year-old. What if we could just stop the increase in the rate of annual probability of dying at age 10, 40, 60 or 80, and the risk remains constant? How would this impact our life expectancy? We should strive for the goal of keeping the risk of dying constant as we age as we work to restore muscle, cognitive, and immune function for those over 50.

As highlighted in the tables below, if our risk of dying stays at a 40-year-old risk level, a 60-year-old would have a life expectancy of ~300 years.

EXPECTED LIFETIME IF THE ANNUAL PROBABILITY OF DYING RATE REMAINS CONSTANT FOR THAT AGE GROUP.

Age Group	Male Expected Remaining Lifetime (years)	Female Expected Remaining Lifetime (years)
10 years	~7,752	~9,709
40 years	~263	~476
60 years	~71	~115
80 years	~15	~22

CURRENT LIFE EXPECTANCY AND EXPECTED LIFE EXTENSION BY AGE GROUP WITHOUT INCREASED RISK

Age Group	Male Death Probability (% per year)	Female Death Probability	Current Life Expectancy (Male)	Current Life Expectancy (Female)	Extended Life Expectancy (Male)	Extended Life Expectancy (Female)
10 years	0.013%	0.01%	74	80	7,826	9,789
40 years	0.38%	0.21%	77	81	340	558
60 years	1.40%	0.87%	80	84	151	199
80 years	6.50%	4.60%	88	89	103	111

Tables calculated with the aid of ChatGPT O1-Preview

Apple co-founder Steve Wozniak once proposed that we will have achieved Artificial General Intelligence (AGI) when a machine can enter a stranger's home and make a pot of coffee. I say AGI will be achieved when it saves 1 million human lives a year. Artificial superintelligence (ASI) will arrive when it can save 1 billion human lives.

What goal was/is more challenging to achieve?

- The goal in 2004 to have 1 billion smartphones in use: achieved in 2011

- The goal in 2024 is to have 1 billion robots in use: predicted in 2033–2035

- Since 2016, 1.2–1.6 billion smart phones shipped per year

- In 2022, 5 billion people were using smartphones.

According to the article "Criteria for AGI/ASI Arrival: A Gradual, Emergent Process Over the Next 20 Years" AI functionality will not happen all at once, but it will be gradual—emergent based on what it can do that is useful and solves problems.

WHY DOES IT MATTER?

Death will come to each of us, but just as humanity has significantly extended life expectancy between 1900 and 2024, by using the tools of AI we can extend healthy life and avoid avoidable death by many years.

CALL TO ACTION

Reduce your chance of avoidable death and maximize your healthy lifespan.

> *"Don't die from something stupid between now and the mid-2030s."*—**Peter H. Diamandis**

Reducing avoidable deaths is a personal and global imperative, and AI is uniquely positioned to play a transformative role. By leveraging AI-driven technologies, individuals can monitor their health in real time, identify early warning signs of disease, and access personalized care plans. AI-powered wearable devices, predictive models, and virtual health assistants offer unprecedented opportunities for proactive health management. Beyond personal care, AI is revolutionizing public health with predictive analytics that allocate resources effectively, optimize vaccination programs, and ensure timely medical interventions. The integration of AI into healthcare promises a future where millions of avoidable deaths are prevented annually.

DOUG HOHULINON USING AI TO SAVE ONE BILLION LIVES

I am working to save one million lives a year in the next decade and 1 billion lives this century by leveraging AI and especially Super Useful AI (SUAI), working on projects where the AI system has to be right: healthcare, road safety/AV, governance/policy, energy, and education. Humanity has made remarkable strides in extending life expectancy and improving the quality of life. Yet, avoidable deaths—those preventable through timely intervention, medical care, or better living conditions—still claim millions of lives annually. These deaths arise from chronic diseases, infectious illnesses, accidents, recklessness and limited medical literacy and healthcare access. As the global population ages and healthcare systems face increasing complexity, addressing these challenges becomes more urgent.

AI is a transformative force in combating preventable deaths. Its ability to analyze vast datasets, identify patterns, and optimize resources can revolutionize preventive medicine, diagnostics, and treatment delivery. It can be a case manager's medical personal assistant to help optimize a person's care. AI-driven tools offer solutions such as predicting disease outbreaks, personalizing treatments, and improving diagnostics. Beyond hospitals, AI addresses social determinants of health, ensuring solutions reach underserved and at-risk populations.

Avoiding preventable deaths requires a multifaceted approach integrating technology, policy, and access to care. AI can expand healthcare access by enhancing diagnostic accuracy and reducing treatment delays. AI-powered telemedicine platforms reach underserved regions, while predictive analytics identify at-risk populations for early intervention. Administrative inefficiencies, particularly in the U.S., can be streamlined through AI automation, reallocating resources to patient care. By embracing AI's potential for efficiency and prevention, the U.S. and other nations can significantly reduce avoidable deaths and improve millions of lives.

Reducing Healthcare Costs with AI

The U.S. healthcare system is projected to reach $7.7 trillion in expenditures by 2032, with its share of GDP rising to 19.7%. Addressing this unsustainable trajectory, AI presents a transformative opportunity to reduce costs without compromising quality or access. (https://www.healthaffairs.org/doi/10.1377/hlthaff.2024.00469).

By targeting inefficiencies across key stakeholder groups, AI could save $200–$360 billion between 2023 and 2028, (http://www.nber.org/papers/w30857).

Key Areas of Savings:

1. **Hospitals:** Hospitals, accounting for the largest share of healthcare costs, could achieve $60–$120 billion in savings (5–11%) through AI-powered automation in administrative tasks such as scheduling, billing, and resource allocation. AI-driven systems can also enhance clinical workflows, reducing waste and improving care delivery.

2. **Physician Groups:** AI tools such as automated patient record management and clinical decision support systems can save physician groups $20–$60 billion (3–8%). These innovations reduce administrative burdens and streamline diagnostics, allowing healthcare providers to focus on patient care.

3. **Private and Public Payers:** AI can deliver $110–$150 billion in savings for payers by improving claims processing, fraud detection, and care

management. Predictive analytics also help identify high-risk populations, enabling proactive interventions that reduce costly hospitalizations.

4. **Other Sites of Care:** AI applications in telemedicine and remote patient monitoring can save $10–$30 billion (1–4%) by extending care access while reducing reliance on expensive in-person visits.

Implementing AI solutions could save $200–$360B in 2023–2028.

Broader Impacts:

AI's ability to automate routine processes, enhance predictive analytics, and optimize resource allocation addresses systemic inefficiencies. Approximately 35% of projected savings come from reducing administrative costs, which consume significant portions of healthcare budgets. Moreover, AI-driven insights can shift the focus toward preventive care, lowering long-term costs associated with chronic disease management. By embracing AI, the U.S. healthcare system can transition toward a more efficient, affordable, and patient-centric model, alleviating economic pressures while improving outcomes.

How AI Can Help Save Lives and Prevent Avoidable Deaths

While technological advances have saved billions of lives, significant disparities in health outcomes remain globally. Cardiovascular diseases, respiratory illnesses, cancers, and infectious diseases continue to claim millions of lives each year. With the integration of AI-driven technologies, there is a profound opportunity to reduce mortality from preventable causes. AI can improve early detection, optimize treatments, monitor chronic conditions, and enhance preventive care, ensuring that healthcare systems become more proactive and available to everyone.

Technological innovations have already saved billions of lives, and AI has the potential to revolutionize healthcare further. As of today, the annual lives saved across various domains of healthcare technologies—such as pharmacology, preventive medicine, and medical procedures—highlight the importance of continuous innovation. With an estimated 68 million deaths occurring globally in 2021, 39 million of these resulting from the top 10 causes of death, AI presents an opportunity to reduce preventable deaths substantially. By 2030, we can anticipate AI contributing to lives saved at an unprecedented scale across critical areas of healthcare. Below are some key areas where AI can intervene:

1. Cardiovascular Diseases (Ischaemic Heart Disease and Stroke)

- Current Deaths: 9.1 million from ischaemic heart disease, 6.6 million from stroke.

- AI Impact by 2030: Early detection, predictive algorithms, and personalized interventions can significantly reduce these deaths. AI systems that monitor heart conditions remotely and alert physicians to early warning signs will become more widespread.

- Estimate of Lives Saved: AI could reduce cardiovascular deaths by 10–15%, translating to 1.5 million lives saved annually by 2030.

2. Respiratory Diseases (COPD and Lower Respiratory Infections)

- Current Deaths: 3.23 million (COPD) and 2.5 million (lower respiratory infections).

- AI Impact by 2030: Smart inhalers and AI-powered spirometry can reduce hospitalizations and manage chronic conditions better. AI tools will also improve disease monitoring and respiratory care in remote areas.

- Estimate of Lives Saved: AI could reduce respiratory-related deaths by 10–12%, resulting in 500,000 lives saved annually.

3. **Cancer Detection and Treatment (Lung, Breast, and Colorectal Cancers)**

- Current Deaths: 1.9 million from trachea, bronchus, and lung cancers.

- AI Impact by 2030: AI-driven diagnostics using imaging tools and genomics-based personalized treatments will become more accurate and accessible. Early detection programs supported by AI can improve survival rates significantly.

- Estimate of Lives Saved: With improved screening, AI could help save 400,000 lives annually by 2030 across various cancer types.

4. **Diabetes Management and Prevention**

- Current Deaths: 2 million, with a rising trend.

- AI Impact by 2030: Predictive models for early diagnosis and AI-powered mobile apps for real-time glucose monitoring can reduce complications. AI could also help detect pre-diabetic conditions and recommend preventive interventions.

- Estimate of Lives Saved: AI could help prevent and manage diabetes effectively, saving 200,000 lives annually by 2030.

5. **Alzheimer's Disease and Chronic Disease Monitoring**

- Current Deaths: 1.8 million from Alzheimer's and dementia.

- AI Impact by 2030: AI tools that track cognitive decline and assist caregivers can slow disease progression and improve outcomes. Better management of chronic diseases using AI could also reduce deaths from other conditions.

- Estimate of Lives Saved: AI could save 100,000 lives annually by improving chronic disease management.

6. Pandemic and Infectious Disease Control

- Current Deaths: 8.8 million from COVID-19 in 2021.

- AI Impact by 2030: AI can enhance pandemic preparedness by predicting outbreaks, optimizing vaccine distribution, and improving early detection through data analysis.

- Estimate of Lives Saved: AI interventions could prevent future pandemic deaths, saving 1 million lives annually by 2030.

7. AI-Driven Preventive Medicine and Public Health Initiatives

- Current Impact: Vaccination programs, anti-smoking campaigns, and other preventive measures save millions of lives.

- AI Impact by 2030: AI-based predictive tools will help allocate resources efficiently, optimize vaccination campaigns, and promote healthy behaviours.

- Estimate of Lives Saved: AI could contribute to an additional 500,000 lives saved annually through preventive medicine.

8. Reducing Medical Errors and Improving Adherence

- Impact of Errors: Medical errors are responsible for significant avoidable deaths globally.

- AI Impact by 2030: AI-powered clinical decision support systems will reduce diagnostic errors, while smart reminders will improve patient treatment adherence.

- Estimate of Lives Saved: Reducing medical errors and improving adherence could save 300,000 lives annually.

By integrating AI into healthcare systems, the potential cumulative impact by 2030 across these key areas can be substantial, as summarized in this table:

Health Area	Potential Lives Saved Annually by 2030 Using AI	Global Mortality: Top Causes (2021)
Cardiovascular Diseases (Heart, Stroke)	1.5 million	Ischaemic heart disease – 9.1 million Stroke – 6.6 million
Respiratory Diseases (COPD, Infections)	500,000	COPD – 3.2 million Lower respiratory infections – 2.5 million
Cancer Detection and Treatment	400,000	9.7 million people died of cancer worldwide. That's 1 out of every six deaths. Over 610,000 cancer deaths happen in the U.S.
Diabetes Management	200,000	Diabetes – 2 million
Alzheimer's and Chronic Disease Monitoring	100,000	Alzheimer's and dementias – 1.8 million
Pandemic and Infectious Disease Control	1 million	COVID-19 – 8.8 million
Preventive Medicine and Public Health	500,000	Lower respiratory infections (infectious diseases)
Reducing Medical Errors and Adherence	300,000	3 million deaths occur annually due to unsafe care - half of this harm is attributed to medications (World Health Organization)

The Lancet article "Ten Americas: A Systematic Analysis of Life Expectancy Disparities in the USA", updates the 2006 Eight Americas study, expanding it to include ten groups based on race, ethnicity, geography, and socioeconomic factors. It highlights growing disparities in life expectancy in the U.S. from 2000 to 2021, with gaps increasing from 12.6 years in 2000 to 20.4 years in 2021. Significant declines occurred during COVID-19, particularly for American Indian and Alaska Native populations. While some groups, notably Black Americans, showed progress pre-2020, systemic inequities and regional inequalities persisted, emphasizing the need for targeted policies addressing health and socioeconomic disparities. (Dwyer-Lindgren, Laura et al. "Ten Americas: a systematic analysis of life expectancy disparities in the USA." *The Lancet*, 404. 10469. 2299-2313).

Here's how AI can support the key calls to action from the paper, along with the potential lives AI could save based on each strategy's impact on reducing life expectancy disparities:

1. Reduce Health Inequities

AI Solution: Use machine learning models to identify and predict systemic health disparities by analyzing socioeconomic, racial, and geographic data patterns. AI could propose interventions, such as improving healthcare access or targeting high-risk groups with specific programs.

Potential Lives Saved: An estimated 100,000–150,000 lives annually if disparities in healthcare access and outcomes are significantly reduced.

2. Focus on Marginalized Groups

AI Solution: Implement AI-driven programs to optimize healthcare delivery to the most disadvantaged groups, such as AIAN populations or low-income rural communities. For example, AI can help triage patients remotely, prioritize care delivery, and tailor educational materials.

Potential Lives Saved: Up to 50,000 lives annually by improving outcomes for the most vulnerable populations.

3. Improve Data Collection

AI Solution: Use AI to automate and enhance the collection and analysis of health data, correcting for biases and inaccuracies in death certificates, population records, and health surveys. AI models can also integrate social determinants of health to improve precision.

Potential Lives Saved: Indirectly supports other initiatives; could enable more targeted actions that save 20,000–40,000 lives annually.

4. Enhance Equity in Healthcare Access

AI Solution: Deploy AI-powered telehealth platforms and diagnostic tools in underserved areas. AI can also optimize transportation and resource allocation for mobile clinics and preventive healthcare drives.

Potential Lives Saved: 80,000–120,000 lives annually through better preventive and primary care access.

5. Invest in Socioeconomic Factors

AI Solution: Leverage AI to identify communities with high socioeconomic risks (e.g., food deserts, poor housing, or low educational attainment) and design intervention programs such as job training or housing subsidies linked to better health outcomes.

Potential Lives Saved: 40,000–60,000 lives annually by addressing upstream determinants of health.

6. Leverage Local and National Planning

AI Solution: Use AI to create detailed, real-time dashboards for policymakers, combining health and socioeconomic data to guide targeted resource allocation and long-term policy initiatives.

Potential Lives Saved: Enables scalable improvements across other interventions, indirectly saving up to 100,000 lives annually.

7. Monitor and Evaluate Progress

AI Solution: Use AI-powered monitoring tools to track health outcomes and assess the effectiveness of interventions over time, identifying gaps and allowing rapid adaptation.

Potential Lives Saved: Indirectly supports life-saving actions across all initiatives, enabling additional savings of 30,000–50,000 lives annually.

8. Address COVID-19 Impacts

AI Solution: AI models can predict COVID-19 outbreaks, optimize vaccine distribution, and analyze the effectiveness of pandemic recovery programs to ensure vulnerable populations recover equitably.

Potential Lives Saved: 50,000–100,000 lives over five years by mitigating long-term health disparities exacerbated by the pandemic.

9. Enhance Preventive Healthcare

AI Solution: Develop AI tools that identify high-risk individuals for chronic diseases and suggest personalized prevention programs, such as health screenings, lifestyle changes, or vaccinations.

Potential Lives Saved: 100,000 lives annually through early intervention.

Cumulative Impact

By implementing AI-driven strategies across these areas, the U.S. could save an estimated 500,000 to 800,000 lives annually by addressing systemic health disparities and achieving better healthcare outcomes for those who need it the most.

AI's Transformative Potential in Reducing Avoidable Deaths

The integration of AI into healthcare has the potential to save millions of lives annually by 2030. With advancements in predictive analytics, personalized treatment, remote monitoring, and preventive healthcare, AI will enhance the efficiency of healthcare delivery. Noncommunicable diseases such as cardiovascular conditions, cancers, and diabetes, which account for the majority of deaths, stand to benefit the most from AI innovations. Additionally, AI will improve healthcare access in underserved areas, mitigate the impact of future pandemics, and reduce avoidable deaths caused by medical errors. Governments, healthcare providers, and technology companies must collaborate to build responsible AI systems prioritizing patient safety, fairness, and transparency. With the right approach, AI can become a cornerstone of public health, ensuring millions more lives are saved in the years to come.

When will AI be able to save more lives than toilets?

AI has the transformative potential to address and reduce the most avoidable causes of death throughout human history, ensuring that the mistakes of the past are not repeated and that preventable deaths in the present are minimized. One of AI's most impactful contributions lies in advancing early detection and prevention of infectious diseases, historically the most significant contributors to human mortality. Through AI-driven surveillance systems, patterns of disease outbreaks can be identified in real-time, enabling rapid containment and resource allocation. Similarly, AI-powered diagnostics can deliver accurate and early detection of diseases such as tuberculosis and malaria, significantly reducing mortality rates.

In maternal and infant health, AI can address the historically high rates of childbirth-related deaths. Predictive analytics and wearable technologies can monitor maternal health in real-time, alerting healthcare providers to complications before they escalate. In regions with limited healthcare infrastructure, AI-powered telemedicine platforms can provide critical guidance to expectant mothers and local practitioners, ensuring safer deliveries. AI also offers solutions for addressing malnutrition and unsafe water, which remain crucial drivers of preventable deaths. Machine learning models can optimize food distribution in famine-stricken areas. At the same time, AI-enabled water quality monitoring systems can ensure access to safe drinking water, preventing diseases such as cholera and dysentery. AI's ability to analyze vast datasets can help policymakers allocate resources effectively, reducing deaths from violence, natural disasters, and environmental exposure. Predictive algorithms can anticipate disaster-prone zones, optimize evacuation plans, and improve early warning systems for earthquakes, floods, and other calamities. By leveraging AI responsibly, humanity can significantly reduce the burden of avoidable deaths, ensuring that future generations can live longer, healthier lives while learning from past lessons.

Solving Global Physical Disability

Over the last 20 years, the global number of Years Lived with Disability (YLD) has increased in absolute terms due to demographic changes and the rising prevalence of non-communicable diseases and injuries. While precise annual figures are scarce, estimates indicate a growth from approximately 760 million YLDs in 2000 to about 850 million YLDs in 2019. This trend underscores the need for global health strategies focusing on prevention, management, and support for chronic health conditions that contribute to long-term disability.

In addressing the global challenge of physical disabilities, AI can emerge as a transformative tool capable of driving meaningful change. By harnessing the power of AI, we can redefine how disabilities are managed, treated, and ultimately mitigated. AI-enabled technologies, such as advanced prosthetics, exoskeletons, and brain-computer interfaces, are already demonstrating their ability to restore mobility and independence for individuals with physical disabilities. These innovations leverage AI to provide real-time adaptability, personalized solutions, and seamless integration with human physiology. Moreover, AI-powered diagnostics and predictive analytics offer new pathways for early detection and prevention of conditions that could lead to long-term disability, reducing the global burden captured by metrics such as Disability-Adjusted Life Years (DALYs). The integration of AI into healthcare systems also fosters accessibility. Telemedicine platforms driven by AI can deliver care to underserved populations, while AI-powered language and assistive tools minimize communication and mobility barriers. By processing vast amounts of data from sources such as The Lancet's Global Burden of Disease studies, AI can help policymakers identify trends and allocate resources effectively, ensuring targeted interventions where they are most needed.

Digital Wellbeing in the Age of AI

We now live in the digital world for 7 hours a day. As we transition into an era where AI is deeply embedded in personal and professional realms, achieving balance and intentionality in our digital interactions is crucial for holistically thriving.

The Six Dimensions of Wellness in a Digital Age

Caitlin Krause's *Digital Wellbeing: Empowering Connection with Wonder and Imagination in the Age of AI* introduces a framework built on six dimensions of wellness: intellectual, occupational, spiritual, social, emotional, and physical. These dimensions emphasize a comprehensive approach to well-being, reminding us that proper health extends beyond the absence of illness to include intellectual growth, meaningful work, spiritual connection, vibrant relationships, emotional regulation, and physical vitality. AI's role in this ecosystem can be transformative if used thoughtfully to enhance these areas.

1. Intellectual Wellness:

AI enables personalized learning experiences tailored to individual interests and goals. Adaptive learning platforms use AI to analyze learning styles, offering customized content that challenges and stimulates intellectual growth. For example, AI tools such as Duolingo for language learning or Coursera's AI-powered recommendations ensure intellectual curiosity is nurtured in engaging, practical ways. The Lancet highlights that addressing the 14 modifiable risk factors collectively could potentially prevent or delay nearly 40–50% of dementia cases, depending on global implementation The lack of good quality education and cognitive stimulation throughout life can contribute to a 7% risk factor.

2. Occupational Wellness:

AI is reshaping the workplace by automating routine tasks, freeing professionals to focus on creative and fulfilling endeavours. It can assist in career planning through personalized skill assessments and training programs, helping individuals align their work with their passions. However, as AI transforms industries, it also raises the need for lifelong learning to adapt to evolving job markets.

3. Spiritual Wellness:

Spiritual well-being is about finding meaning and connection. AI tools help individuals cultivate mindfulness and appreciate life's depth, fostering a sense of peace and purpose. From a religious perspective, these tools can deepen one's relationship with God, providing access to scripture space for reflection, prayer, and discernment. By enabling practices such as guided meditations on scripture or virtual faith communities, AI can support believers in seeking spiritual growth and aligning their lives with God's purpose.

4. Social Wellness:

Loneliness and social isolation are significant challenges in the digital age. During the COVID-19 pandemic, many lacked "someone to love, something to do, and something to hope for." AI can bridge these gaps through platforms that facilitate authentic connections. For instance, AI moderates online communities, suggesting meaningful interactions based on shared interests, and dating apps powered by AI have helped 61% of couples meet, reducing barriers to finding companionship.

5. Emotional Wellness:

AI-driven mental health tools offer significant potential in supporting emotional well-being. Chatbots such as Woebot provide on-demand counselling, teaching users strategies to manage stress and anxiety. By offering accessible and affordable mental health support, AI helps individuals navigate their emotions effectively.

6. Physical Wellness:

Wearable technologies and AI-driven health platforms are pivotal in enhancing physical wellness. Fitbit and Apple Watch monitor heart rate, sleep patterns, and activity levels, providing actionable insights to encourage healthier habits. AI systems also predict potential health risks, empowering users to take preventive measures.

Source: National Wellness Institute's six dimensions of wellness, Six Dimensions Overview_Introduction & Summary_2023

The Double-Edged Sword of AI in Social Interactions

AI's potential to enhance digital well-being is immense, but it also carries risks. As Hank Green wrote, "We seek the safety of isolation even as it kills us." AI can exacerbate artificial intimacy if not designed with care, replacing authentic relationships with shallow digital interactions. Social media platforms, for example, often prioritize engagement metrics over meaningful connections, leading to feelings of alienation despite increased online activity.

Addressing these concerns requires an intentional design prioritizing meaningful interactions over superficial engagement. For example, AI could be harnessed to foster genuine community building by recommending group activities or facilitating deeper discussions within online spaces. Similarly, initiatives integrating AI into offline community engagement, such as neighbourhood wellness programs, could bridge the gap between the digital and physical worlds.

AI as a Catalyst for Hope and Connection

AI's ability to inspire hope is perhaps its most transformative potential. By connecting individuals with opportunities for personal growth, meaningful work, and supportive communities, AI can help people find a sense of purpose. For instance, platforms that connect volunteers to causes they care about, powered by

AI-matching algorithms, can enable people to contribute to society while fostering a sense of belonging.

During crises such as the COVID-19 pandemic, AI was invaluable in mitigating loneliness and facilitating connection. Virtual reality (VR) platforms powered by AI allow families and friends to share experiences despite physical separation. These technologies illustrate how AI can bring people together, even across vast distances, creating shared joy and support opportunities.

Educating for Responsible AI Use

To maximize the benefits of AI while minimizing its harms, digital literacy and education are paramount. Teaching individuals how to use AI tools intentionally can empower them to make choices that prioritize their well-being. For example, setting boundaries around screen time, practicing mindfulness in digital spaces, and seeking AI tools that align with personal wellness goals can help users navigate the digital landscape more effectively.

Future policies and regulations should also focus on ethical AI design to ensure systems promote well-being. By embedding principles of transparency, fairness, and inclusivity into AI development, society can create tools that uplift rather than exploit.

A Future of Digital Flourishing

As Krause suggests, digital well-being is not about rejecting technology but integrating it thoughtfully into our lives. When designed and used responsibly, AI has the power to not only enhance individual wellness but also foster collective flourishing. It can inspire creativity, strengthen connections, and offer hope in ways previously unimaginable.

The challenge lies in harnessing this potential with intention, ensuring that the digital tools we create and use align with our deepest values. By doing so, we can transform the digital world into a space of wonder, imagination, and well-being—a place where humanity thrives in harmony with technology.

If you want to age twice your age, go on bed rest.

As told in the article, "Exercise and Aging Can you walk away from Father Time" at www.health.harvard.edu, "In 1966, five healthy men volunteered for a research study ... all they had to do was spend three weeks ... resting in bed. But it probably didn't seem so good when they got out of bed at the end of the trial. Testing the men before and after exercise, the researchers found devastating changes that included faster-resting heart rates, higher systolic blood pressures, a drop in the heart's maximum pumping capacity, a rise in body fat, and a fall in muscle strength. In just three weeks, these 20-year-olds developed many physiologic characteristics of men twice their age."

Fortunately, "they put the men on an 8-week exercise program. Exercise did more than reverse the deterioration by bed rest since some measurements were better than ever after the training."

This underscores a crucial lesson: regular exercise is vital to maintaining and improving physical health. Inactivity accelerates aging—just 3 weeks of bed rest could set someone back decades in health metrics. Exercise, however, is the ultimate antidote, keeping the body strong and resilient at any age. If you do not use it, you will lose it.

The Risk of AI

Unfortunately, AI can be used to kill people as well. Ukraine can produce 2 million drones by the end of 2024 (uatv.ua)

In the report "Global Catastrophic Risk Assessment" (RAND, Chapter 9 Artificial Intelligence: Summary of Risk) it is noted that "the OECD has defined AI system as a machine-based system that, for explicit or implicit objectives, infers from the input it receives how to generate outputs (e.g., predictions, content, recommendations, or decisions that influence physical or virtual environments)." As AI systems grow more powerful and take on more significant roles, rapid development and deployment raise concerns about potential accidental and intentional risks.

"Potential factors that contribute to risk:

- AI can supply enabling information to malicious actors.
- AI systems are prone to specification and robustness failures.
- AI could open the door to adversarial attacks.
- AI errors can be difficult to detect and correct."

AI amplifies risks such as nuclear conflict, pandemics, climate change, and gradual disruptions in governance, economies, and infrastructure. Mitigating these risks requires designing safer systems, ensuring responsible deployment, and maintaining continuous oversight.

Will AI and technology save or kill more people in the 21st century? It is up to all of us to create safe technology that saves lives. I hope for a 21st century of abundance, but that will take hard work, and I will not just throw AI and technology at the problem.

WHEN WILL AI AND TECHNOLOGY SAVE ONE BILLION LIVES IN THE 21ST CENTURY?

In the ongoing narrative of technological evolution, AI and related technologies are poised to reduce mortality rates worldwide dramatically. By analyzing historical precedents and current technological trajectories, AI could save millions and tens of millions of lives annually and, eventually, one billion lives within this century.

Ultimately, the question remains: Will AI and technology save more lives than they end? This balance will depend heavily on innovation, investment, ethical implementations, global regulations, and public trust in AI systems. Responsible development and deployment of AI, guided by ethical standards and robust governance, will be crucial in ensuring that the benefits of AI in life preservation vastly outweigh any potential harm. This approach aligns with the humanities goal of leveraging technology to innovate and significantly enhance the quality and longevity of human life.

Our collective challenge extends beyond milestones as we chart our course through the 21st century. It encompasses a broader duty to support and nurture all 10 billion humans who will call our planet home in this century. This immense responsibility demands a multifaceted approach, grounded in sustainability, innovation, and inclusivity, to forge a world that not only welcomes new lives with boundless opportunities but also respects and cares for those nearing the end of their journey. Innovation in education and healthcare stands as a pillar for building this future. Universal access to quality education, enabled by technological advancements, can unlock the potential of billions, equipping them with the knowledge and skills to tackle the challenges of tomorrow. Similarly, healthcare innovation must aim for universality and accessibility, ensuring that every individual, irrespective of where they are born, has the chance to lead a healthy and fulfilling life. Digital health platforms, telemedicine, and breakthroughs in medical science offer promising paths to this end.

Supporting all 10 billion humans by the end of this century requires a concerted effort from governments, businesses, civil society, and individuals. It calls for a shift in mindset from short-term gains to long-term focus, from individual success to collective well-being. Our legacy should be defined not just by the technological marvels we create or the wealth we accumulate but by our ability to make a sustainable world that nurtures and promises a better tomorrow for all its inhabitants.

Conclusion: The Future is Ours to Shape

As we stand on the brink of 2030, the pace of transformation around us is both exhilarating and humbling. The technologies explored in this book—AI, robotics, data centers, semiconductors, XR, blockchain, healthcare, wellness, synthetic biology, and more—are not just tools of progress; they are the keystones of a new era. These innovations can multiply human potential, enabling us to tackle some of the most pressing challenges in history—including saving 1 billion lives this century—and unlocking opportunities previously confined to the realm of imagination.

Technology is not a replacement for human ingenuity—it is a force multiplier. With AI, we can accelerate breakthroughs in medicine, education, and climate solutions. Robotics can free us from monotonous tasks, allowing us to focus on creativity and empathy. Data centers are the engines powering our digital future, enabling the seamless flow of information that connects the world. Each of these technologies, when wielded responsibly, can uplift individuals, amplify creativity, and expand the boundaries of what's possible.

This transformation isn't without its challenges. The same technologies that promise abundance and progress can deepen inequalities if left unchecked. Imagine a future where AI and robotics improve lives for the privileged few but leave billions behind. This is a scenario we must avoid. Our task is not just to innovate but to ensure that the benefits of these innovations are shared equitably. We must build systems that democratize access to tools, knowledge, and opportunities so that everyone can contribute to and benefit from this new world.

Imagine a world where a teenager in rural India designs a renewable energy solution using AI tools available through the cloud, or an entrepreneur in Nairobi collaborates with a partner in Stockholm to create a low-cost desalination system for water-scarce regions. Technology can make this world a reality, but only if we lay the foundation today. By removing barriers to entrepreneurship, enabling universal internet access, and fostering an inclusive digital economy, we can create a generation of problem-solvers equipped to tackle local and global challenges.

The future doesn't just happen—it is built by the decisions we make today. Governments must prioritize investments in sustainable infrastructure and equitable policies. Businesses must embrace ethical innovation, ensuring their technologies serve humanity, not just the bottom line. Educators and institutions must prepare a workforce that is adaptable, resilient, and capable of thriving in an era of rapid change.

As individuals, we have a role, too. Whether advocating for ethical AI, supporting local entrepreneurs, or learning how to use these technologies ourselves, we must actively participate in shaping the future. This is not the time for passive spectatorship. The exponential era requires us all to be bold, curious and engaged.

The future is a garden waiting to be cultivated. Technology provides the seeds of possibility, but we can only ensure it blossoms for all through equitable distribution and ethical responsibility. Some parts of the garden will flourish without care while others remain barren—a stark reminder of what's at stake.

We are at a crossroads. One path leads to a Star Trek future—an abundant, collaborative, and hopeful world where technology uplifts every member of society. The other leads to a Star Wars future marred by division, exploitation, and conflict. The choice is ours to make. We can build a future where humanity thrives together by harnessing the extraordinary tools at our disposal with care, empathy, and foresight.

This book is more than a snapshot of technological progress; it is a blueprint for what we can achieve together. Let it inspire you to act—not just for personal gain, but for collective progress. The challenges before us are vast, but so is our potential. If we seize this moment with purpose and unity, we can create a future where technology catalyzes prosperity, equity, and human flourishing.

The future isn't written yet. Let's make it extraordinary—for us and for future generations.

Alan Smithson, Doug Hohulin & Harvey Castro, MD

Appendix

Podcasts

Delve deeper with our All AI cloned voices podcast: The GPT Podcast on Spotify and theGPTPodcast.com. https://www.thegptpodcast.com.

Alan's podcast "XR For Business" interviews those making and using XR technology for business applications and shares their challenges and successes. Search "XR For Business" wherever you get your podcasts. Julie Smithson has a sister podcast, "XR For Learning" exploring the myriad ways XR technology can be deployed for learning and education. Search "XR For Learning" wherever you get your podcasts.

Books by Dr. Harvey Castro

Expand your horizons with Dr Harvey Castro's enlightening publications. Search "Harvey Castro, MD" on Amazon or visit his Author Page for an exhaustive list. Below is a partial list of the books.

Success Reinvention: Unveil the journey of a preeminent thought leader. Dr. Castro's unique blend of entrepreneurship, medical expertise, and visionary leadership offers insights to surmount life's challenges and realize your aspirations.

ChatGPT and Healthcare: Unlocking The Potential Of Patient Empowerment: A seminal exploration of artificial intelligence's role in elevating patient care. This comprehensive guide discusses ChatGPT technology, its patient benefits, and its transformative impact on healthcare.

Revolutionize Your Health and Fitness with ChatGPT's Modern Weight Loss Hacks: My Personal Experience of Hacking My Body for Weight Loss: Chronicles an individual's tech-assisted weight loss journey, providing a deep dive into the science, techniques, benefits, and risks of body hacking.

The AI-Driven Entrepreneur: A blueprint for leveraging artificial intelligence in entrepreneurial ventures. Discover how AI can revolutionize your business, from market research to product development.

Solving Infamous Cases with Artificial Intelligence: Investigates ChatGPT-4's role in redefining criminology, offering new insights into forensic science, crime scene analysis, and the resolution of historical crimes.

Illness and Infamy: Explores the impact of medical conditions on historical figures and notorious criminals, providing a unique perspective on the interplay between health, power, and infamy.

Featured Collections and Tools:

Entrepreneur GPT Collection: Tailored for entrepreneurs, this collection equips you with innovative tools and strategies. It includes Navigator AI for personalized advice, Success Reinvention Workstation for growth, BIZ AI Guru for intelligence, and AI Catalyst for innovation.

Literary GPT: Offers classic literature resources and Navigator AI for literary analysis, Detective Hub for mystery insights, and Historical Minds AI for educational content on literary figures.

Fitness & Longevity GPT: Provides AI-curated wellness advice, Navigator AI for personalized guidance, Fit Tech Advisory for technology reviews, and Longevity Learner AI for life-quality education.

Healthcare & Medical Education GPT: "The Healthcare & Medical Education GPT collection features Navigator AI for healthcare knowledge, Healthtech Innovator for technology insights, History Health AI for tracking medical milestones, and Medi Assistant for providing guidelines to medical staff.

Spiritual GPT & Translator GPT Español: Delve into spiritual texts with Bible Insight & Koran Insight AI GPT, and enhance communication in Spanish and Portuguese with AI Comunicador.

Access the **Entire GPT Collection** with a free 7-day trial, ideal for extensive learning across various interests, leveraging AI for a comprehensive experience.

Convo with GPT Website

Visit Convo with GPT:

https://www.convowithgpt.com

Embark on a journey of AI-integrated learning with Convo with GPT, a platform designed to immerse you in a broad spectrum of disciplines, including healthcare, entrepreneurship, literature, and beyond. This platform utilizes advanced AI to offer immersive, personalized educational experiences, enhancing your understanding and engagement in each subject area.

Further Reading

Principles for Dealing with the Changing World Order: Why Nations Succeed and Fail Ray Dalio

The Sixth Extinction: An Unnatural History Elizabeth Kolbert

This Changes Everything: Capitalism vs. The Climate Naomi Klein

Power And Prediction: The Disruptive Economics of Artificial Intelligence Avi Goldfarb, Ajay Agrawal, and Joshua Gans

Economics in America: An Immigrant Economist Explores the Land of Inequality Angus Deaton

Autonomy Lawrence D. Burns

High and Mighty: The Dangerous Rise of the SUV Keith Bradsher

Introduction to Autonomous Mobile Robots Roland Siegwart and Illah Reza Nourbakhsh

Autonomous Vehicles: Opportunities, Strategies, and Disruptions Hannibal Travis

The Coming Wave Mustapha Suleyman

Superintelligence: Paths, Dangers, Strategies Nick Bostrom

The Alignment Problem: Machine Learning and Human Values Brian Christian

Life 3.0: Being Human in the Age of Artificial Intelligence Max Tegmark

AI Superpowers: China, Silicon Valley, and the New World Order Kai-Fu Lee

Quantum Computing for Everyone Chris Bernhardt

Dancing with Qubits Robert S. Sutor

Chip War: The Fight for the World's Most Critical Technology Chris Miller

The Age of Surveillance Capitalism: The Fight for a Human Future at the New Frontier of Power Shoshana Zuboff

The New Economics of Semiconductors: Moore's Law and Technological Change in the Data Center Industry - Ashish Arora, Sharon Belenzon, and Andrea Patacconi

The Innovators: How a Group of Hackers, Geniuses, and Geeks Created the Digital Revolution Walter Isaacson

Data Center Handbook Hwaiyu Geng

Made to Measure: New Materials for the 21st Century Philip Ball

Stuff Matters: The Strange Stories of the Marvellous Materials that Shape Our Man-made World Mark Miodownik

A Materials Science Guide to Superconductors and How to Make Them Super Susannah Speller

The 3D Printing Handbook Ben Redwood, Filemon Schöffer, and Brian Garret

First Layer: The Beginner's Guide to 3D Printing Pete Stagman

Functional Design for 3D Printing Clifford T Smyth

The Future of Architecture in 100 Buildings Marc Kushner

Building a Better World: 50 Ideas for a Sustainable Future Richard Rogers

Design Like You Give a Damn: Architectural Responses to Humanitarian Crises Cameron Sinclair, Kate Stohr, and Samuel Mockbee

The Lean Builder: A Builder's Guide to Lean Construction Joe Donarumo and Keyan Zandy

3D Printing for Architects and Designers Michael D. Hensel

Synthetic Biology: A Primer Paul S. Freemont and Richard I. Kitney

Synthetic Biology: Tools for Engineering Biological Systems Daniel G. Gibson and Clyde A. Hutchison III

Regenesis: How Synthetic Biology Will Reinvent Nature and Ourselves George Church and Ed Regis

Synthetic Biology: A Lab Manual Josefine Liljeruhm, Erik Gullberg, and Anthony C. Forster

Synthetic Biology: From iGEM to the Artificial Cell Markus Schmidt

The History of the Future: Oculus, Facebook, and the Revolution That Swept Virtual Reality Blake J. Harris

Dawn of the New Everything: A Journey Through Virtual Reality Jaron Lanier

Experience on Demand: What Virtual Reality Is, How It Works, and What It Can Do Jeremy Bailenson

The Metaverse: And How it Will Revolutionize Everything Matthew Ball

Snow Crash Neal Stephenson (where the word Metaverse comes from)

A Brief History of Video Games: From Atari to Virtual Reality Richard Stanton

Charlie Fink's Metaverse: An AR Enabled Guide to AR & VR Charlie Fink (Note: Alan Smithson has a chapter in this book)

Our Next Reality: How the AI-Powered Metaverse Will Reshape the World Alvin W. Graylin and Louis Rosenberg

Interconnected Realities: How the Metaverse Will Transform Our Relationship with Technology Forever Leslie Shannon

Step into the Metaverse Dr. Mark Van Rijmenam

The Metaverse: A Professional Guide Tom Ffiske

Into the Metaverse: The Essential Guide to the Business Opportunities of the Web3 Era Cathy Hackl

The Metaverse Handbook Quharrison Terry and Scott "DJ Skee" Keeney

Brain-Computer Interfacing: An Introduction Rajesh P. N. Rao

"How ARK Is Thinking About Humanoid Robotics" (ark-invest.com)

"The rise of the robots: charted"
 (www.ft.com/content/fcc917ab-5b12-4222-901a-6922f1d6894a)

Ion Propulsion: The Future of Space Travel James W. Smith

Physics of the Future: How Science Will Shape Human Destiny and Our Daily Lives by the Year 2100 Michio Kaku

The Case for Space: How the Revolution in Spaceflight Opens Up a Future of Limitless Possibilities Robert Zubrin

The Future of Humanity: Terraforming Mars, Interstellar Travel, Immortality, and Our Destiny Beyond Earth Michio Kaku

Space 2.0: How Private Spaceflight, a Resurgent NASA, and International Partners are Creating a New Space Age Rod Pyle

Space Tourism: A New Frontier The International Space University

AI and the Future of Education: Teaching in the Age of Artificial Intelligence Priten Shah

Augmented Reality and the Future of Education Technology Rashmi Aggarwal et al.

A Brief History of the Future of Education Ian Jukes & Ryan L. Schaaf

The Google Infused Classroom Holly Clark and Tanya Avrith

Instructional Technology Theory in the Post-Pandemic Era David D. Carbonara and Lawrence A. Tomei

Blog: https://educatorsinvr.com/

Parent & Student Resource for VR: https://medium.com/vr-ar-parent-student -resource

Teaching 2030: What We Must Do for Our Students and Our Public Schools Barnett Berry, et al.

The Condition of the Working Class in England Friedrich Engels

Independent People Halldór Laxness

Taking Up Space: The Black Girl's Manifesto for Change Chelsea Kwakye and Ore Ogunbiyi

Admissions Mira Harrison

Scavengers Darren Simpson

2030: How Today's Biggest Trends Will Collide and Reshape the Future of Everything Mauro F. Guillén

Clean Disruption of Energy and Transportation Tony Seba

Mobility 2040: Exploring Emerging Trends Radically Transforming Transportation Systems in the US Galo Bowen

Faster, Smarter, Greener: The Future of the Car and Urban Mobility Venkat Sumantran et al.

The Future of Mobility: Scenarios for the United States in 2030 Johanna Zmud et al.

The Alignment Problem Brian Christian

The Hundred-Page Machine Learning Book Andriy Burkov

The No-Nonsense, Deep Dive into LLM Prompt Engineering for People Who Don't Know Sh*t About It Timothy E. Bates

The Idea Factory John Gertner

The Makers of The Microchip Christophe Lécuyer and David Brock

The Man Behind the Microchip Leslie Berlin

Physics of Semiconductor Devices S.M. Sze

Semiconductor Device Fundamentals Robert F. Pierret

Cisco Data Center Fundamentals Somit Maloo

Enterprise Data Center Design and Methodology Rob Snevely

The Datacenter as a Computer: Designing Warehouse-Scale Machines Luiz André Barroso, Urs Hölzle, and Parthasarathy Ranganathan

Building a Modern Data Center: Principles and Strategies of Design Scott D. Lowe

Where wizards stay up late Katie Hafner

The Art of Invisibility Kevin Mitnick

Data and Goliath: The Hidden Battles to Collect Your Data and Control Your World Bruce Schneier

Cybersecurity for Dummies Joseph Steinberg

Permanent Record Edward Snowden

Hacking: The Art of Exploitation Jon Erickson

Privacy is Power Carissa Veliz

The Fifth Domain Richard A. Clarke and Robert K. Knake

Sandworm: A New Era of Cyberwar and the Hunt for the Kremlin's Most Dangerous Hackers Andy Greenberg

Countdown to Zero Day Kim Zetter

Material Science and Engineering: An Introduction William Callister and David Rethwisch

The New Science of Strong Materials JE Gordon

Stuff: The Materials the World is Made of Ivan Amato

Elements of Solid State Physics Michael Rudden and John Wilson

3D Printing for Dummies Richard Horne

Designing 3D Printers Neil Rosenberg

3D Scanner Unabridged Guide Mark Koch

The First Layer: The Beginner's Guide to 3D Printing Joseph Larson

Innovations, Disruptions and Future Trends in the Global Construction Industry Temitope Omotayo et al.

Future Homes Images Publishing Group

Green Building Illustrated Francis D.K. Ching and Ian M. Shapiro

Building Reuse: Sustainability, Preservation, and the Value of Design
Kathryn Rogers

Strategies for Sustainable Architecture Paola Sassi

Housing: Strategies for Urban Living Various Authors

Green Buildings Pay Brian W. Edwards

The Construction Technology Handbook Hugh Seaton

Missing Middle Housing: Thinking Big and Building Small to Respond to Today's Housing Crisis Daniel Parolek

The Second Machine Age Erik Brynjolfsson and Andrew McAfee

Rise of the Robots: Technology and the Threat of a Jobless Future Martin Ford

Automation and the Future of Work Aaron Benanav

The Future of Work: Robots, AI, and Automation Darrell M. West

AI-Savvy Leader: Nine Ways to Take Back Control and Make AI Work
David De Cremer

The Skill Code: How to Save Human Ability in an Age of Intelligent Machines
Matt Beane

Our Final Invention: Artificial Intelligence and the End of the Human Era
James Barrat

The VR Book: Human-Centered Design for Virtual Reality Jason Jerald

Performing Mixed Reality Steve Benford and Gabriella Giannachi

The Immersive Classroom Jaime Donally

Spatial Computing Shashi Shekhar

Reality Check: How Immersive Technologies Can Transform Your Business
Jeremy Dalton

The Augmented Workforce: How Artificial Intelligence, Augmented Reality, and 5G Will Impact Every Dollar You Make Cathy Hackl and John Buzzell

The Metaverse: And How It Will Revolutionize Everything Matthew Ball

The Fourth Transformation: How Augmented Reality & Artificial Intelligence Will Change Everything Robert Scoble and Shel Israel

Learning Virtual Reality: Developing Immersive Experiences and Applications for Desktop, Web, and Mobile Tony Parisi

UX for XR: User Experience Design and Strategies for Immersive Technologies Cornel Hillmann

Hyper-Reality: The Art of Designing Impossible Experiences Keiichi Matsuda

Virtual Natives: How a New Generation is Revolutionizing the Future of Work, Play, and Culture Leslie Shannon and Catherine D. Henry

Delta-V Daniel Suarez

The Space Barons Christian Davenport

Beyond: Our Future in Space Chris Impey

The Martian Andy Weir

Escaping Gravity Lori Garver

Packing for Mars Mary Roach

Voyage Stephen Baxter

Critical Mass Daniel Suarez

The Mission David W. Brown

Mining the Sky John S. Lewis

Project Hail Mary Andy Weir

Asteroid Mining 101 John S. Lewis

A History of Global Health Randall Packard

The Emperor of All Maladies: A Biography of Cancer Siddhartha Mukherjee

The Ghost Map: The Story of London's Most Terrifying Epidemic Steven Johnson

And the Band Played On: Politics, People, and the AIDS Epidemic Randy Shilts

The Great Influenza: The Story of the Deadliest Pandemic in History John M. Barry

The Immortal Life of Henrietta Lacks Rebecca Skloot

Mountains Beyond Mountains: The Quest of Dr. Paul Farmer, a Man Who Would Cure the World Tracy Kidder

Plagues and Peoples William H. McNeill

The Cambridge World History of Human Disease Kenneth F. Kiple

Digital Wellbeing Caitlin Krause

Scary Smart: The Future of Artificial Intelligence and How You Can Save Our World Mo Gawdat

The Coming Wave: Technology, Power, and the Twenty-first Century's Greatest Dilemma Mustafa Suleyman

Superconvergence: How the Genetics, Biotech, and AI Revolutions Will Transform Our Lives, Work, and World Jamie Metzl

Human + Machine: Reimagining Work in the Age of AI Paul R. Daugherty and H. James Wilson

AI 2041: Ten Visions for Our Future Kai-Fu Lee and Chen Qiufan

The Future of Humanity: Terraforming Mars, Interstellar Travel, Immortality, and Our Destiny Beyond Earth Michio Kaku

AI Superpowers: China, Silicon Valley, and the New World Order Kai-Fu Lee

Our Final Invention: Artificial Intelligence and the End of the Human Era James Barrat

Think and Grow Rich Napoleon Hill

As a Man Thinketh James Allen

The Power of the Subconscious Mind Joseph Murphy

Abundance: The Future Is Better Than You Think Peter Diamandis and Steven Kotler

The Magic of Thinking Big David J. Schwartz

About the Authors

Alan Smithson, B.Sc. is a globally renowned futurist, keynote speaker, and author dedicated to inspiring socially, economically, and environmentally responsible action. A father, founder, and DJ, Alan makes complex topics engaging and actionable, sharing insights in his TEDx talk, The Marriage of Education and Technology.

As co-founder of Emulator (touchscreen DJ technology), MetaVRse (3D creation platform), Unlimited Awesome (education for entrepreneurs), and Clean Data Centers (nuclear-powered data centers), Alan leads innovations in AI, blockchain, metaverse, education, and clean energy. He advises HSTAR Space, invests in Lawfully Minded (AI legal resource), and has been a Techstars mentor, SXSW Pitch judge, and One Creativity Awards juror.

Named a Top 50 Digital Futurist, Alan has spoken at CES, SXSW, TEDx, and AWE, with features in *Forbes*, *HBR*, and *CNN*. As DJ "Lord Alan," he's performed globally and won DJ Mag's Innovative Product award. Learn more at AlanSmithson.com.

Doug Hohulin, B.Sc.EE @ Purdue University is a futurist and strategic thought leader with a deep background in smart cities, communications, AI, Digital Health, and other emerging technologies. With 22 years at Motorola and 11 years at Nokia in engineering, account management, business development, and strategy, his career spanned industries such as telecommunications, and digital health. As a thought leader, Doug specializes in developing and deploying cutting-edge technologies such as 5G, and autonomous vehicles. His work bridges the gap between technical implementation and strategic vision, helping businesses, governments, and communities design future-ready blueprints for exponential leap.

Connect with Doug: https://www.linkedin.com/in/doughohulin/

Together, Alan, Doug, and Harvey bring a wealth of knowledge, experience, and visionary thinking to *2030: A Blueprint for Humanity's Exponential Leap*. Their combined expertise spans technology, urban development, healthcare, entertainment, and education, creating a comprehensive roadmap for understanding and thriving in the rapidly changing world ahead.

Harvey Castro, MD, MBA, is a distinguished physician, healthcare consultant, author, and innovator committed to advancing digital health awareness and transforming healthcare. With a diverse career in leadership, technology, and entrepreneurship, Dr. Castro is dedicated to sharing his insights through consulting, writing, and public speaking. He actively seeks to contribute his expertise to corporate boards and mentorship programs. Learn more about Dr. Castro's work and mission at HarveyCastroMD.com

Connect with Dr. Harvey Castro

LinkedIn: @HarveyCastroMD

Twitter: @HarveyCastroMD

Facebook: @HarveyCastroMD

Instagram: @harveycastromd

YouTube: @HarveyCastroMD